Solid Mechanics and Its Applications

Volume 245

Aims and Scope of the Series

The fundamental questions arising in mechanics are: *Why*?, *How*?, and *How much*? The aim of this series is to provide lucid accounts written by authoritative researchers giving vision and insight in answering these questions on the subject of mechanics as it relates to solids.

The scope of the series covers the entire spectrum of solid mechanics. Thus it includes the foundation of mechanics; variational formulations; computational mechanics; statics, kinematics and dynamics of rigid and elastic bodies: vibrations of solids and structures; dynamical systems and chaos; the theories of elasticity, plasticity and viscoelasticity; composite materials; rods, beams, shells and membranes; structural control and stability; soils, rocks and geomechanics; fracture; tribology; experimental mechanics; biomechanics and machine design.

The median level of presentation is to the first year graduate student. Some texts are monographs defining the current state of the field; others are accessible to final year undergraduates; but essentially the emphasis is on readability and clarity.

More information about this series at http://www.springer.com/series/6557

Peter Davies · Yapa D.S. Rajapakse
Editors

Durability of Composites in a Marine Environment 2

Springer

Editors
Peter Davies
Marine Structures Laboratory
IFREMER
Plouzané
France

Yapa D.S. Rajapakse
Solid Mechanics Division
Office of Naval Research
Arlington, VA
USA

ISSN 0925-0042 ISSN 2214-7764 (electronic)
Solid Mechanics and Its Applications
ISBN 978-3-319-65144-6 ISBN 978-3-319-65145-3 (eBook)
https://doi.org/10.1007/978-3-319-65145-3

Library of Congress Control Number: 2017947858

Printed on acid-free paper

This Springer imprint is published by Springer Nature
The registered company is Springer International Publishing AG
The registered company address is: Gewerbestrasse 11, 6330 Cham, Switzerland

The original version of the book frontmatter was revised: Series volume number has been changed. The erratum to the book frontmatter is available at https://doi.org/10.1007/978-3-319-65145-3_13

Contents

Introduction

Following the successful IFREMER/ONR Workshop on "Durability of Composites in a Marine Environment", held at the IFREMER Centre in Nantes, France, in August 2012 [1], a second Workshop was organized at the IFREMER Centre in Brest, France, in 2016, with about 50 participants from eight countries, and from academia, research institutions and industry. The main objective of both Workshops, and of this and the previous book, is to provide a state-of-the-art overview of current research on the behaviour of composite materials in a marine environment. This has been achieved by bringing together leading researchers in the field, providing invited lectures and discussions over two days, and then inviting all presenters to prepare a chapter for this book.

While fibre-reinforced polymer composites have a long history in marine applications (pleasure boats, military ships, submarines), durability aspects are rarely explicitly integrated in design. There have been some attempts to address this in European projects and within the ONR Solid Mechanics program, but we are only starting to include key aspects such as coupling between water ingress and mechanical behaviour. This is a new research area, made possible by the development of computing power; it requires a close collaboration between researchers with experience of materials in a marine environment and modellers developing predictive tools, in order to provide naval architects with validated long-term behaviour models.

In the 4 years since the first Workshop, there have been some significant developments in the use of composites. The offshore industry is now starting to realise the weight gain potential of high-performance composites in particular for flexible risers. A second emerging area is renewable marine energy, with considerable progress towards commercial exploitation of both floating wind turbines and tidal turbines. For both offshore and marine energy structures, long-term durability of composites is a major issue, and this was acknowledged by inviting speakers from both of these industries.

This book is structured in a similar way to the Workshop. The first day focussed on understanding and modelling water effects and degradation; the second day was oriented more towards applications. The first chapter by Colin describes work to improve our understanding of water diffusion mechanisms (Chapter "Nonempirical

Kinetic Modelling of Non-fickian Water Absorption Induced by a Chemical Reaction in Epoxy-Amine Networks"). This is followed by two chapters describing the influence of water on interfaces in composites, first from Peters who describes the ageing of fibre coatings (Chapter "Influence of Glass Fibre Sizing and Storage Conditions on Composite Properties"), then from Carlsson and Du who show how water migrates along fibre/matrix interfaces (Chapter "Water Uptake in Polymer Composites with Voids"). Penumadu then provides an overview of ageing, from fibres through to naval composites and sandwich materials (Chapter "Durability of US Naval Composites and Sandwich Structures: Science Framework Considering Multi-scale Response in Harsh Sea Environment"). Miyano and colleagues describe the extension of their time-temperature accelerating method to predict tensile creep behaviour (Chapter "Statistical Long-Term Creep Failure Time of Unidirectional CFRP"). This is followed by two contributions on modelling, first from Clément and colleagues who describe a multi-physics approach to model diffusion-mechanical property coupling (Chapter "Multiphysics Modeling of the Hygro-Mechanical Behavior of Heterogeneous Materials"). Then Sobey and colleagues apply reliability based modelling to marine composites (Chapter "Reliability of Composite Marine Structures"). An overview by Echtermeyer et al. of a conceptual framework to integrate durability modelling from the molecular level through to structures completes this first part of the book (Chapter "Multiscale Modelling of Environmental Degradation—First Steps").

The chapters based on the second day of the Workshop begin with an overview of composite possibilities for Offshore Oil and Gas by Melot (Chapter "Present and Future Composites Requirements for the Offshore Oil and Gas Industry"). This is followed by two chapters on the use of composites for tidal turbine blades, first by Dawson et al, from a blade producer's viewpoint (Chapter "Composite Materials in Tidal Energy Blades"), followed by Bradaigh et al who describe blade design (Chapter "Influence of Composite Fatigue Properties on Marine Tidal Turbine Blade Design"). The book finishes with a look at the durability of a range of potential materials for future marine applications, with reduced environmental impact (Chapter "Marine Ageing Behaviour of New Environmentally-Friendly Composites").

The chairmen would like to thank IFREMER and ONR for support, and to acknowledge the enthusiastic help of members of the local organizing committee (Pierre-Yves Le Gac, Maelenn Le Gall, Dominique Choqueuse), and the invaluable secretarial assistance of Ann-Sophie Delmaire in setting up and ensuring the smooth running of the Workshop. Thanks also to all those who reviewed the papers for this publication, and to all participants of the Workshop.

Reference

1. Davies P, Rajapakse YDS (eds) (2013) Durability of composites in a marine environment. Springer

Nonempirical Kinetic Modeling of Non-fickian Water Absorption Induced by a Chemical Reaction in Epoxy-Amine Networks

Xavier Colin

Abstract In the last two decades, several studies made in our laboratory have shown that hydrolytic reactions may occur during water absorption and may be responsible for behavioral deviations from the classical Fick's law in epoxy-amine networks. On one hand, water is chemically consumed by specific groups initially present in the repetitive structural unit (e.g., unreacted epoxies and amides) or formed by oxidation under operating conditions. On the other hand, water establishes strong molecular interactions (hydrogen bonds) with new highly polar groups resulting from hydrolysis (alcohols and acids). Due to both contributions, the kinetic curves of water absorption no longer tend towards an equilibrium value, i.e., a final saturation plateau, but display a slow and continuous increase over time in the water mass uptake. On this basis, a diffusion/reaction model has been developed for predicting such a peculiar water sorption behavior. In addition, the classical Henry's law has been modified for describing the changes in boundary conditions during the course of the hydrolytic reaction. This chapter provides an overview of the recent theoretical advances made in this field and demonstrates, through two case studies, the good predictive value of the kinetic modeling approach set up in our laboratory.

Keywords Epoxy-amine networks · Water absorption · Hydrolysis · Fick's law deviation · Kinetic modeling

1 Introduction

Epoxy-amine networks are widely used in many industrial fields, such as aeronautics and space, automotive industry, nuclear energy, building and civil engineering, as matrices of fiber reinforced composite materials, structural adhesives of

X. Colin (✉)
Laboratoire PIMM, Arts et Métiers ParisTech, 151 Boulevard de L'Hôpital,
75013 Paris, France
e-mail: xavier.colin@ensam.eu

P. Davies and Y.D.S. Rajapakse (eds.), *Durability of Composites in a Marine Environment 2*, Solid Mechanics and Its Applications 245,
https://doi.org/10.1007/978-3-319-65145-3_1

bonded assemblies, but also protective coatings, in particular paintings, varnishes and lacquers. Many studies have been conducted for tentatively elucidating their physical and chemical aging mechanisms and for evaluating their long-term durability under the combined effects of various environmental stresses such as temperature, UV and ionizing radiations, oxygen, water, etc. [1–3].

These polymers are known to be very sensitive to moisture. First of all, they can absorb large amounts of water due to the presence of highly polar groups (hydroxyls for example) in their repetitive structural unit, which establish strong molecular interactions (hydrogen bonds) with water molecules. That is the reason why they undergo an important external plasticization when they are initially in the glassy state. As an example, an ideal DGEBA-DDM network absorbs more than 2 wt% of water in 100% RH at 25 °C, which reduces its glass transition temperature T_g by about 36 °C [4].

In addition, epoxy-amine networks can undergo a hydrolysis reaction when they contain specific chemical groups known to be reactive with water. These latter can be initially present in the repetitive structural unit as, for instance, amides in networks based on amidoamine hardeners [5] or unreacted epoxies in non-ideal networks [6]. But, they may also be formed by a radical chain oxidation reaction activated by external factors (e.g., increase in temperature, UV or ionizing radiation, or chemical reagent acting as a radical initiator) under operating conditions. These chemical groups are then amides and esters resulting from the oxidation of amino- and oxy-methylenes respectively [7]. In both cases, hydrolysis will lead to the formation and accumulation of new hydroxyl groups (alcohols and acids) and consequently, to significant changes in water transport properties in epoxy-amine networks.

This chapter proposes to review all these physico-chemical mechanisms and to derive, from these latter, a general kinetic model capable of predicting two examples of behavioral deviations from the classical Fick's law reported in the literature for epoxy-amine networks.

2 Water Absorption in the Absence of Hydrolysable Groups

In the absence of hydrolysable groups in epoxy-amine networks, water absorption obeys the classical Fick's second law [6, 8, 9]. A typical kinetic curve of water absorption is plotted versus the square root of time in Fig. 1a. Two time domains can be clearly distinguished:

I. The transient regime during which the water mass uptake m (expressed in wt %) increases linearly with the square root of time. The slope of this straight line allows the coefficient D of water diffusion to be determined. During this period, water is heterogeneously distributed in the sample thickness as schematized in Fig. 1b.

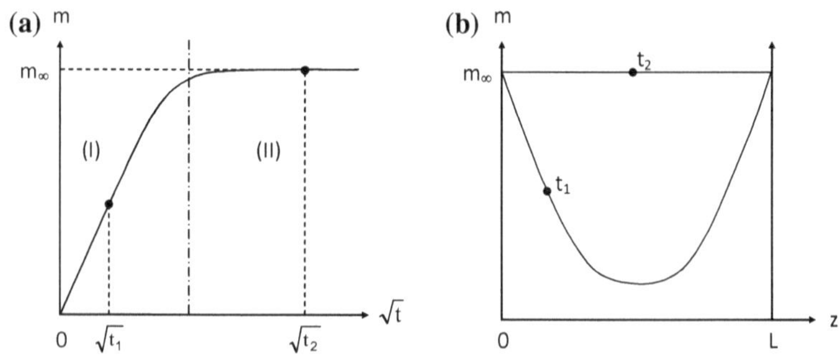

Fig. 1 **a** Typical kinetic curve of water absorption for a Fick's diffusion process. **b** Water gradient in the sample thickness at two different times of exposure

II. The final saturation plateau for which m has reached its maximum value m_∞ throughout the sample thickness (for the exposure conditions under study). Thenceforth, water is homogeneously distributed throughout the sample thickness.

The equilibrium mass uptake m_∞, mass fraction μ_∞ and concentration C_∞ of absorbed water in the polymer sample are linked by the following relationships:

$$\mu_\infty = \frac{m_\infty}{1 + m_\infty} \quad \text{and} \quad C_\infty = \frac{m_\infty}{18} \ (\text{expressed in mol g}^{-1}) \tag{2.1}$$

2.1 *Equilibrium Properties*

Since water is far below its critical point (T_C = 647 K), the pertinent environmental parameter is the water activity *a*. Let us remember that for humid atmospheres:

$$a = \frac{p_V}{p_{VS}} = \frac{RH}{100}, \tag{2.2}$$

where p_V and p_{VS} are respectively the water partial pressure in the atmosphere and the saturated vapor pressure of water.

In the absence of hydrolysable groups in epoxy-amine networks, C_∞ varies linearly with a over the whole activity range (Fig. 2). In other words, C_∞ obeys the classical Henry's law [10]:

$$C_\infty = H\, a, \tag{2.3}$$

where *H* is almost independent of temperature.

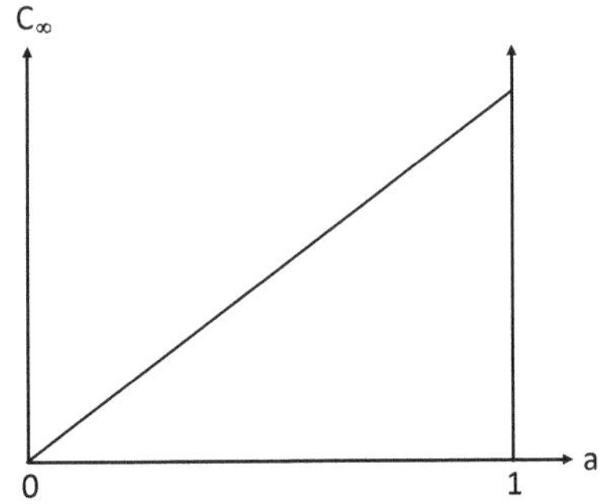

Fig. 2 Typical sorption isotherm of Henry's type

2.2 *Origin of Hydrophilicity*

In the 1980s, many authors have tried to relate the polymer hydrophilicity to the available free volume [11–14], but such theory fails in explaining why free-volume-rich organic substances, such as liquid hydrocarbons and silicone rubbers, are hydrophobic. Today, it seems clear that hydrophilicity is essentially linked to the molecular interactions between water molecules and polar groups in the polymer matrix. In a first approach, m_∞ can be roughly estimated from Hildebrandt's solubility parameter δ. Indeed this latter, which is directly related to the chemical structure of the polymer, gives access to polymer-solvent interactions. It can be written:

$$\delta = \frac{F}{V} \tag{2.4}$$

with F the molar attraction constant and V the molar volume of the polymer.

According to Van Krevelen's theory [15], both quantities are molar additive functions, i.e., they can be calculated by summing the molar contributions of the different chemical groups composing the repetitive structural unit:

$$F = \sum F_i \quad \text{and} \quad V = \sum V_i \tag{2.5}$$

Table 1 summarizes the values of F_i and V_i proposed by Small [16], Van Krevelen [17], Hoy [18] and Fedors [19] for usual chemical groups.

The literature values of m_∞ in 100% RH at 25 °C of different polymers have been compiled and plotted versus δ (calculated from Eqs. 2.4 and 2.5) in the log–log diagram of Fig. 3.

A correlation is clearly observed between both quantities. It can be written:

$$m_\infty = 7.1 \times 10^{-10}\,\delta^{6.86} \tag{2.6}$$

Table 1 Molar contributions to F [16–18] and V [19] of usual chemical groups. Ph designates an aromatic ring

Group	F_i ($J^{1/2}$ $cm^{3/2}$ mol^{-1})	V_i (cm^3 mol^{-1})
$-CH_3$	370	33.5
$-CH_2-$	275	16.1
>CH–	115	−1
>C<	15	−19.5
–Ph–	1680	89.4
–O–	200	3.8
>N–	125	−9
$-NH_2$	490	19.2
–CO–	560	10.8
–O–CO–	590	18
–O–CO–O–	760	22
–CO–O–CO–	960	30
–NH–CO–	1110	9.5
–CO–OH	825	28.5
–OH	610	10
$-SO_2-$	845	19.6

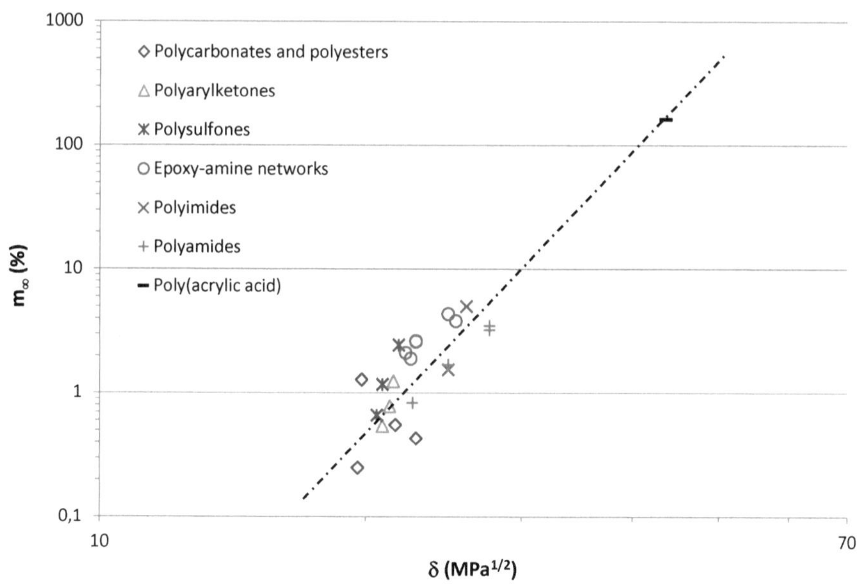

Fig. 3 Equilibrium water mass uptake (in 100% RH at 25 °C) versus solubility parameter for 24 polymers

Epoxy-amine networks are moderately polar polymers ($22 < \delta < 25$ MPa$^{1/2}$). They absorb typically between 1 and 4 wt% of water in 100% RH at 25 °C.

2.3 *Effect of Structure on Hydrophilicity*

According to Van Krevelen [15], C_∞ would be a molar additive function, i.e., each chemical group of the repetitive structural unit would be characterized by a molar contribution to hydrophilicity H_i (in mole of water per mole of group) independent of its chemical neighbors:

$$C_\infty = \left(\frac{18}{M_{RSU}}\sum H_i\right) \times \frac{HR}{100}, \tag{2.7}$$

where M_{RSU} is the molar mass of the repetitive structural unit. Table 2 summarizes the values of H_i determined by Barrie [20] for usual chemical groups in 100% RH at 25 °C. On this basis, three main categories of chemical groups can be defined:

- Apolar groups, such as aromatic and aliphatic hydrocarbon structures, which have a negligible contribution to hydrophilicity: $H_i \approx 0$ mol/mol;
- Moderately polar groups, for instance ethers, ketones and esters, which have a relatively low contribution to hydrophilicity: $H_i = 0.1{-}0.3$ mol/mol;
- And highly polar groups, for instance acids, amides, and alcohols, which establish intense molecular interactions with water molecules (hydrogen bonds) and absorb typically $H_i = 1{-}2$ mol/mol.

Equation 2.7 can be used for estimating the orders of magnitude of C_∞ for the main polymer families. In contrast, this approach fails in predicting the hydrophilicity variations within a same polymer family which displays large concentration differences in polar groups, such as epoxy-amine networks [6] and polysulfones [21]. Indeed in these two latter, it has been found that C_∞ does not

Table 2 Molar contributions to hydrophilicity (in moles of water per mole of group) of usual chemical groups in 100% RH at 25 °C [20]. Ph designates an aromatic ring

Polarity	Group	H_i (mol/mol)
– –	$-CH_3$	0.00005
	$-CH_2-$	
	>CH–	
	–Ph–	
+	–O–	0.1
	–CO–	0.2
	–O–CO–	0.3
+++	–CO–OH	1.3
	–CO–NH–	2
	–OH	2

increase linearly but exponentially with the concentration of polar groups (i.e., hydroxyls and sulfones, respectively).

Gaudichet-Maurin et al. [21] have proposed a theory for tentatively explaining such a behavior. According to these authors, an hydrophilic site would not be a single but a pair of polar groups having to satisfy some geometric requirements (in particular an optimal distance between them) in order to establish a double hydrogen bond with a water molecule. As schematized in Fig. 4, the probability of finding two polar groups at this optimal distance would increase exponentially with their concentration, which could effectively explain the mathematical shape observed experimentally for the curve: $C_\infty = f$ ([Polar group]).

More recently, Courvoisier [22] has compiled the literature values of C_∞ in 100% RH at 25 °C of different polymers and plotted them versus the concentration of the most polar group in the repetitive structural unit (Fig. 5). It can be seen that all the points are placed around only three master curves, which allows to distinguish three main categories of polar groups: slightly (esters and carbonates), moderately (aryl ketones, amides and imides), and highly polar groups (hydroxyls and sulfones). Thus, it appears clearly that all carbonyl groups do not display the same behavior. The lower contribution to hydrophilicity of esters and carbonates had been already reported in the literature by several other authors, for instance in reference [23].

Since these three curves display an exponential shape, it can be assumed that the previous theory can be generalized to all polymers. A general semi-empirical structure/C_∞ relationship can be proposed:

$$C_\infty = A\,\mathrm{Exp}\{B[\mathrm{Polar\ group}]\} \times \frac{HR}{100}, \tag{2.8}$$

where A is almost independent of the chemical structure: $A = (1.23 \pm 0.41)\ 10^{-1}$ mol.L^{-1} for all polymers. In contrast, B is an increasing function of the group polarity.

In fact, B is found to be proportional to the "elemental" solubility parameter of the polar group under consideration:

$$B = (6.7 \pm 1.0)\ 10^{-3} \times \delta \tag{2.9}$$

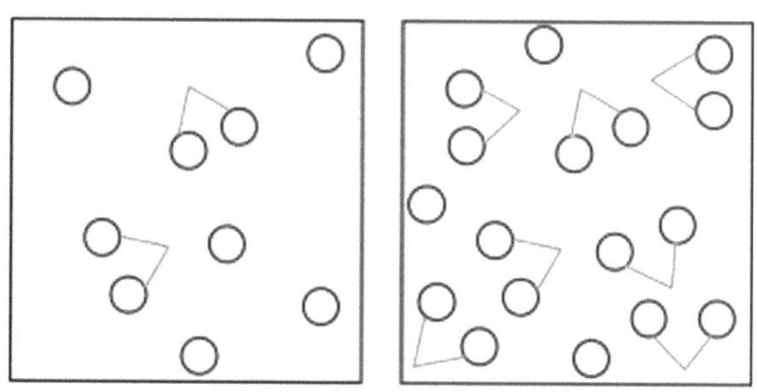

Fig. 4 Schematization of the probability increase of finding two polar groups at an optimal distance for establishing a double hydrogen bond with a water molecule o: polar group. ∧: water molecule

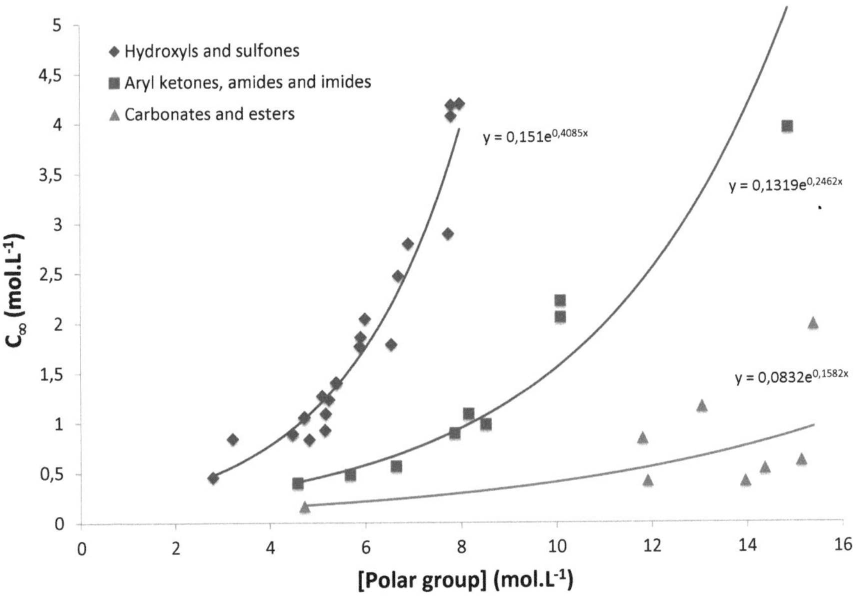

Fig. 5 Equilibrium water concentration (in 100% RH at 25 °C) versus polar group concentration for 40 polymers

Table 3 Solubility parameters of common chemical groups. Ph designates an aromatic ring

Polarity	Group	δ ($MPa^{1/2}$)
– –	$-CH_3$	16.8 ± 2.1
	$-CH_2-$	
	>CH–	
	–Ph–	
+	–O–	22.8
	–O–CO–	27.4 ± 0.1
	–CO–O–CO–	
++	–CO–	32.6 ± 2.3
	–CO–H	
	$-CO-NH_2$	
	–CO–NH–	
+++	$-SO_2$	50.4 ± 2.5
	–OH	

Table 3 summarizes the values of δ calculated for usual chemical groups from the values of cohesive energy E_{coh} and molar volume V, reported respectively by Fedors [19] and Hoy [18], and applying the well-known relationship:

$$\delta = \left(\frac{E_{coh}}{V}\right)^{1/2} \quad (2.10)$$

2.4 Effect of Structure on Diffusivity

In contrast, the relationships between polymer structure and water diffusivity are far from being totally elucidated. Thominette et al. [24] have observed that *D* is a decreasing function of hydrophilicity in different polymer families, for instance in epoxy-amine networks, aromatic polysulfones, and polyester-styrene networks, but also in a series of polyethylenes differing by their initial concentration of oxygenated groups (pre-oxidation). This dependence shows that the diffusion rate is slowed down by the molecular interactions between water molecules and polar groups in the polymer matrix. A three-step scheme has been proposed [6] to describe such a diffusion mechanism (Fig. 6). According to this scheme, water diffusion is not highly influenced by the macromolecular mobility if the dissociation of the water-polymer complex (I) is slower than the water migration between neighboring hydrophilic sites (II).

Courvoisier [22] has compiled the literature values of water permeability *P* in 100% RH at 25 °C of different polymers and plotted them versus the concentration of the most polar group in the repetitive structural unit (Fig. 7). It can be seen that *P* is of the same order of magnitude for all polymers. The data scattering within a same polymer family can be explained by the small effect on *P* of secondary structural factors at higher scales, e.g., differences in crosslinking density in networks, differences in crystallinity ratio and crystalline lamellae thickness in semi-crystalline polymers, etc., which modify the tortuosity of the water diffusion path. In a first approximation, it can be written:

$$P = C_\infty \times D = 10^{-9.7\pm0.7} \quad (\text{expressed in mol m}^{-1}\,\text{s}^{-1}) \quad (2.11)$$

In other words, *D* is inversely proportional to C_∞.

$$[S_1 \ldots H_2O] \rightarrow S_1 + H_2O \quad (I)$$
$$H_2O \rightarrow \text{migration until } S_2 \quad (II)$$
$$S_2 + H_2O \rightarrow [S_2 \ldots H_2O] \quad (III)$$

Fig. 6 Three-step mechanism for water diffusion into polymer matrices [6]. S_1 and S_2 designate two neighboring hydrophilic sites and $[S_i \ldots H_2O]$ a hydrogen bonded water-polymer complex

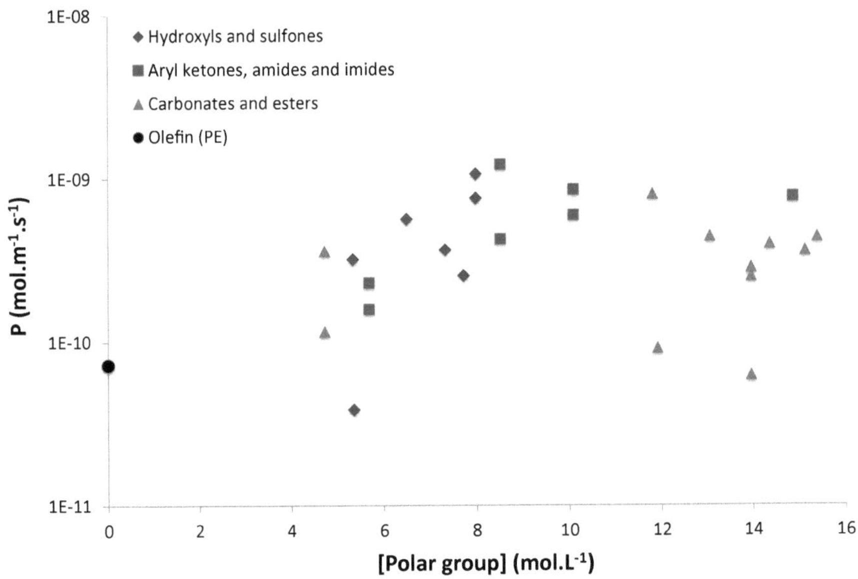

Fig. 7 Water permeability (in 100% RH at 25 °C) versus polar group concentration for 27 polymers

Fig. 8 Typical kinetic curves of water absorption for a Langmuir's diffusion process

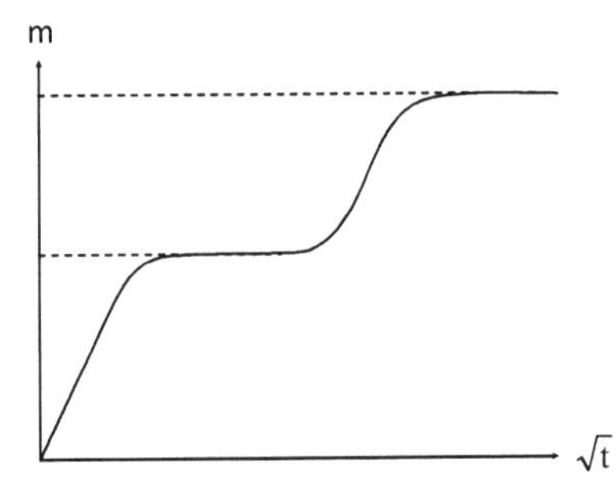

3 Water Absorption in the Presence of Hydrolysable Groups

In the presence of hydrolysable groups in epoxy-amine networks, different behavioral deviations from the classical Fick's law have been reported in the literature. Some of them are clearly related to chemical modifications.

In the case of non-ideal networks containing unreacted epoxy groups, water absorption displays a two-step behavior (Fig. 8) often classified in the categories of

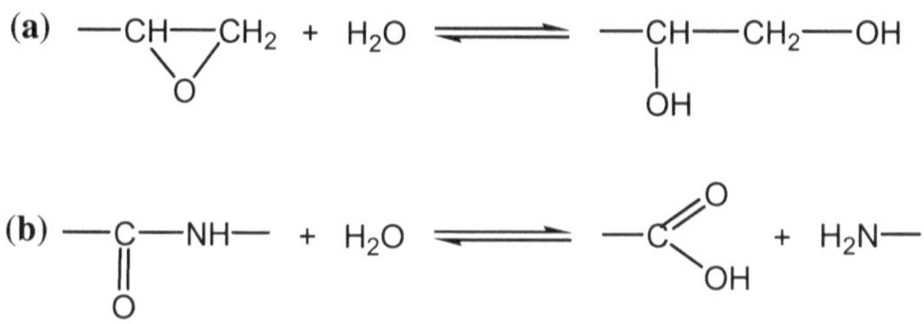

Fig. 9 Hydrolysis of epoxy (**a**) and amide groups (**b**). The double arrows indicate that the reaction is reversible, i.e. equilibrated by the reverse reaction of condensation

Langmuir processes [25, 26]. In this case, water diffusion coexists with a water trapping–untrapping mechanism on particular sites. In fact, this mechanism is simply the reversible hydrolysis of unreacted epoxies [6] as schematized in Fig. 9a. During the second absorption stage characterizing the Langmuir process, the conversion of epoxies into much more polar groups (diols) consumes water molecules, but also increases significantly the polymer hydrophilicity. Of course, both additional contributions have to be considered in any approach of nonempirical kinetic modeling.

More recently, the same type of behavioral deviation has been reported for an ideal epoxy-amidoamine network [27]. However, in contrast to unreacted epoxies, amide groups are not located at dangling chain extremities, but within elastically active chains. As a result, their hydrolysis (Fig. 9b) leads also to a chain scission and, when two hydrolysis events take place on the same elastically active chain (i.e., at high conversion ratios), to the liberation of short macromolecular fragments (diacids) from the epoxy network. The superimposition of the resulting mass loss to the water mass uptake generates a complex sorption behavior [27]. Such behavior was already shown for aliphatic polyamides, in particular for PA 11 [28].

Let us note that hydrolysable groups are not necessarily initially present in epoxy-amine networks. Indeed, it cannot be excluded that they are formed and accumulated by a radical chain oxidation reaction activated by external factors (e.g., increase in temperature, UV or ionizing radiation, or chemical reagent acting as a radical initiator) under operating conditions. In that case, they may be esters and amides resulting from the oxidation of oxy- and amino-methylenes, respectively [7].

4 Proposal of Kinetic Model

On this basis, a general diffusion/reaction model has been developed for tentatively predicting the resulting kinetic curves of water absorption in epoxy-amine networks. Schematically, the water gradient in the thickness L of a semi-infinite plate is calculated from a balanced equation, expressing that the change over time *t* in water concentration *C* in an elementary sublayer located at the depth *z* beneath the sample surface is the difference between the water supply by diffusion (according to the Fick's second law) and the water consumption by the chemical reaction:

$$\frac{\partial C}{\partial t} = D\frac{\partial^2 C}{\partial z^2} - r(C) \tag{4.1}$$

The mathematical expression of the water consumption rate is deduced from a kinetic analysis of the reversible hydrolysis scheme:

$$\begin{array}{ll} E + W \rightarrow A_1 + A_2 & (k_h) \\ A_1 + A_2 \rightarrow E + W & (k_c) \end{array},$$

where E, W, A_1, and A_2 account respectively for hydrolysable groups, water, and hydrolysis products. k_h and k_c are the respective rate constants of the hydrolysis and condensation reactions. The water consumption rate can be written as:

$$r(C) = -\frac{dC}{dt} = k_h[E]\,C - k_c[A_1]\,[A_2] \tag{4.2}$$

The corresponding gradients in the sample thickness of hydrolysable groups and hydrolysis products are determined from complementary equations:

$$\frac{d[E]}{dt} = -k_h[E]\,C + k_c[A_1]\,[A_2] \tag{4.3}$$

$$\frac{d[A_1]}{dt} = \frac{d[A_2]}{dt} = k_h[E]\,C - k_c[A_1]\,[A_2] \tag{4.4}$$

The system of Eqs. 4.1–4.4 has been solved simultaneously in space (*z*) and time (*t*) using the numerical algorithms recommended for problems of chemical kinetics in the literature. Let us note that these algorithms are already implemented in the commercial MATLAB® software under the names of ODE23 s and ODE23 tb. The initial (throughout the sample thickness) and boundary conditions (at the sample surface) used for this resolution are recalled below:

- At $t = 0$ and any z: $C(z, 0) = 0$, $[E](z, 0) = [E]_0$ and $[A_1](z, 0) = [A_2](z, 0) = 0$.
- When $t > 0$, at $z = 0$ and L: $C(0, t) = C(L, t) = C_S(t)$.

Let us remember that, due to the hydrolysis reaction, the water concentration C_S at the sample surface is no longer a constant but it is an increasing function of time. It can be determined from Eq. 2.8 if the total concentration of highly polar groups (i.e., hydroxyls) at any time t is known. The initial value of C_S corresponds to the equilibrium water concentration C_∞ that should be found if the epoxy-amine network under consideration did not contain hydrolysable groups. The increase in C_S against time of exposure can be determined by introducing the solution of Eq. 4.4

into Eq. 2.8. Indeed, for the two hydrolysis examples reported in Fig. 9, this latter writes:

$$C_S = A\,\mathrm{Exp}\{B\,[OH]\} \quad (4.5)$$

with $A = 1.51 \times 10^{-1}$ mol L^{-1} and $B = 4.09 \times 10^{-1}$ L mol^{-1}. If considering the definition of C_∞, this equation can be rewritten:

$$C_S = C_\infty\,\mathrm{Exp}\{B([OH] - [OH]_0)\} \quad (4.6)$$

Since hydroxyl groups (OH) come from alcohols (Al) and acids (Ac), it comes finally:

$$C_S = C_\infty\,\mathrm{Exp}\{B([Al] - [Al]_0)\} \times \mathrm{Exp}\{B([Ac] - [Ac]_0)\} \quad (4.7)$$

In addition, always due to the hydrolysis reaction, the coefficient of water diffusion *D* through the sample thickness is no longer a constant but a decreasing function of time. It can be determined from Eq. 2.11 through replacing C_∞ by C_S. The initial value of *D* corresponds to the coefficient of water diffusion D_0 that should be found if the epoxy-amine network under consideration did not contain hydrolysable groups. Since the water permeability is independent of the concentration of polar groups (Eq. 2.11), it can be written:

$$C_\infty \times D_0 = C_S \times D \quad (4.8)$$

If considering Eq. 4.7, it comes finally:

$$D = \frac{D_0}{\mathrm{Exp}\{B([Al] - [Al]_0)\} \times \mathrm{Exp}\{B([Ac] - [Ac]_0)\}} \quad (4.9)$$

As said before, the local value of the water mass uptake at the depth *z* beneath the sample surface is the sum of two contributions: water diffusion and water consumption:

$$dm(z,t) = \frac{18}{\rho}\left(C + \int_0^t r(C)\,dt\right) \quad (\text{expressed in } \%) \quad (4.10)$$

where ρ is the initial polymer density. The global value of the water mass uptake is simply deduced by summing the local values of dm throughout the thickness *L* of the semi-infinite plate:

$$m(t) = \frac{1}{L}\int_0^L dm(z,t)\,dz \quad (\text{expressed in } \%) \quad (4.11)$$

5 Application to Two Case Studies

The validity of the kinetic model described in the previous paragraph has been checked for the two hydrolysis examples reported in Fig. 9. Experimental data have been recovered from two distinct research studies [6, 27].

In the case of non-ideal networks containing unreacted epoxy groups, hydrolysis products are diols, therefore $[A_1] = [A_2] = [Al]$ and $[Ac] = [Ac]_0 = 0$. As an example, the simulations of the kinetic curves of water absorption in 100% RH at 50 °C of DGEBD-ETHA networks only differing by their initial concentrations of unreacted epoxy groups are reported in Fig. 10. These networks were synthetized from DGEBD/ETHA mixtures with different r = amine/epoxy functional ratios [6].

A satisfactory agreement can be observed between theory and experiment. The values of the rate constants k_h and k_c and the coefficient of water diffusion D_0 used for these simulations are specified in Table 4.

On one hand, it is checked that the rate constants take almost the same values in all simulations. At 50 °C, their average values are:

$$k_h = 3.4 \times 10^{-8}\,\mathrm{L\,mol^{-1}\,s^{-1}} \quad \text{and} \quad k_c = 1.5 \times 10^{-8}\,\mathrm{L\,mol^{-1}\,s^{-1}}$$

Thus, it is found that: $k_h \approx 2 \times k_c$. In other words, hydrolysis of epoxy groups is favored over diols condensation for long periods of exposure.

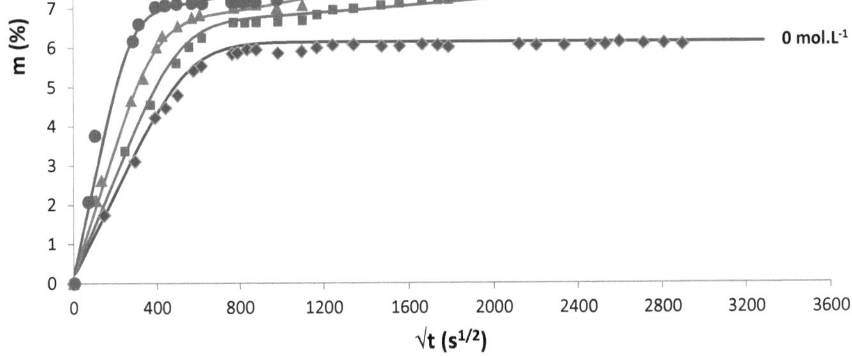

Fig. 10 Kinetic curves of water absorption in 100% RH at 50 °C of non-ideal DGEBD-ETHA networks. Comparison between theory (*solid curves*) and experiments (*symbols*). The concentrations of unreacted epoxies are mentioned next to the curves

Table 4 Rate constants and coefficient of water diffusion used for simulating the kinetic curves of Fig. 10

$[E]_0$ (mol L^{-1})	k_h (L.mol^{-1} s^{-1})	k_c (L mol^{-1} s^{-1})	D_0 (m^2 s^{-1})
0.0	–	–	2.5×10^{-12}
0.8	3.0×10^{-8}	1.9×10^{-8}	3.0×10^{-12}
1.7	3.8×10^{-8}	1.1×10^{-8}	5.0×10^{-12}
2.7	3.3×10^{-8}	1.5×10^{-8}	1.0×10^{-11}

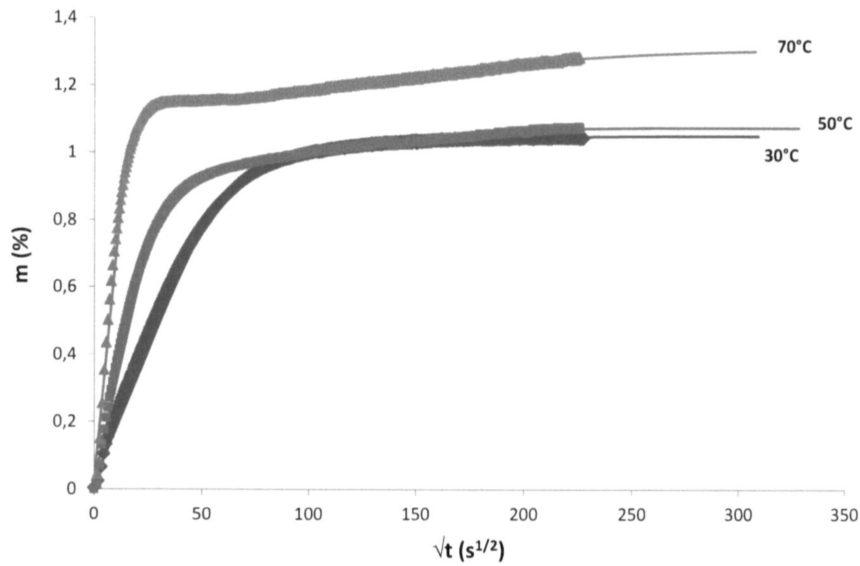

Fig. 11 Kinetic curves of water absorption in 50% RH at 30, 50 and 70 °C of an ideal DGEBA-PAA network. Comparison between theory (*solid curves*) and experiments (*symbols*). The temperatures of exposure are mentioned next to the curves

On the other hand, as expected, the coefficient of water diffusion is a decreasing function of the network polarity. Indeed, D_0 is equal to 2.5×10^{-12} m^2 s^{-1} for the ideal network with an initial concentration of hydroxyl groups of $[OH]_0 =$ 8.0 mol L^{-1}, whereas it increases up to 1.0×10^{-11} m^2 s^{-1} for the least ideal network having an initial concentration of hydroxyl groups of $[OH]_0 = 6.2$ mol L^{-1}.

In the case of ideal networks containing amide groups, hydrolysis products are acids and amines, therefore $[A_1] = [Ac]$, $[A_2] = [Am]$, and $[Al] = [Al]_0$. As an example, the simulations of the kinetic curves of water absorption in 50% RH at 30, 50, and 70 °C of a DGEBA-PAA network are reported in Fig. 11. PAA is a complex amidoamine hardener resulting from the condensation of fatty acids with polyamines (mostly triethylenetetramine) [27].

Table 5 Rate constants and coefficient of water diffusion used for simulating the kinetic curves of Fig. 11

T (°C)	k_h (L mol^{-1} s^{-1})	k_c (L mol^{-1} s^{-1})	D_0 (m^2 s^{-1})
30	6.5×10^{-6}	1.7×10^{-4}	5.0×10^{-13}
50	9.5×10^{-6}	1.5×10^{-4}	2.0×10^{-12}
70	9.5×10^{-6}	1.0×10^{-4}	6.0×10^{-12}

Here also, a satisfactory agreement can be observed between theory and experiment. The values of the rate constants k_h and k_c and the diffusion coefficient D_0 used for these simulations are specified in Table 5.

On one hand, it can be seen that the rate constants are almost independent of temperature. In fact, k_c displays the same order of magnitude as in aliphatic polyamides, e.g., in PA66 [29]. In contrast, k_h is about two decades higher in the epoxy-amidoamine network under consideration than in PA 66. Unfortunately, structure/rate constants relationships are not well known yet in this domain. It is thus too premature to give an explanation.

On the other hand, the values of the coefficient of water diffusion are in very good agreement with those usually reported in the literature for ideal epoxy-amine networks [30]. It is found that D_0 obeys an Arrhenius law with a pre-exponential factor of 1.5×10^{-3} m^2 s^{-1} and an activation energy of 54.9 kJ mol^{-1}.

6 Conclusions

A nonempirical kinetic model has been elaborated for predicting the non-Fickian water absorption induced by a reversible hydrolysis reaction in epoxy-amine networks. This model couples water diffusion and its consumption by the chemical reaction in a system of three differential equations, and takes into account the changes in network hydrophilicity due to the formation of new highly polar groups (alcohols and acids). Its validity has been successfully checked from the experimental results of two case studies conducted in the last two decades in our laboratory.

This model gives access to kinetic parameters (rate constants of hydrolysis and condensation, and coefficient of water diffusion) that are hardly accessible experimentally. The values found for the coefficient of water diffusion are in very good agreement with those already reported in the literature for ideal epoxy-amine networks. In contrast, the values obtained for the rate constants are still questionable. No doubt, these latter have to be the object of a specific study devoted to the humid aging of a large series of epoxy-amine networks containing a well-known concentration of hydrolysable groups after synthesis, but also after storage at room temperature (in particular, in a completely dry atmosphere).

References

1. Colin X, Verdu J (2011) Thermooxidative and thermohydrolytic aging of organic composite matrices. In: Song DB (ed) Resin composites: properties, production and applications, Chap 7. Nova Science Publishers, New York, pp 255–298
2. Colin X, Verdu J (2012) Aging of organic matrix composite materials. In: Nicolais L, Borzacchiello A, Lee Song SM (eds) Wiley Encyclopedia of composites, 2nd edn, Chap 4. Wiley, New York, pp 35–49
3. Colin X, Verdu J (2014) Humid ageing of organic matrix composites. In: Davies P, Rajapakse YDS (eds) Solid mechanics and its applications 208: durability of composites in a marine environment, Chap 3. Springer, Dordrecht, pp 47–114
4. Bellenger V, Verdu J, Morel E (1989) Structure-properties relationships for densely crosslinked epoxy-amine systems based on epoxide or amine mixtures. Part 2: Water absorption and diffusion. J Mater Sci 24:63–68
5. Abdelkader AF, White JR (2005) Water absorption in epoxy resins: The effects of the crosslinking agent and curing temperature. J Appl Polym Sci 98(6):2544–2549
6. Tcharkhtchi A, Bronnec Y, Verdu J (2000) Water absorption characteristics of diglycidylether of butane diol-3,5-diethyl-2,4-diaminotoluene networks. Polymer 41(15):5777–5785
7. Colin X, Fayolle B, Cinquin J (2016) Further progress in the kinetic modelling of the thermal oxidation of epoxy-diamine matrices. Mater Tech 104(2):202
8. Li L, Yu Y, Wu Q, Zhan G, Li S (2009) Effect of chemical structure on water sorption of amine-cured epoxy resins. Corros Sci 51:3000–3006
9. Perrin FX, Nguyen MH, Vernet JL (2009) Water transport in epoxy-aliphatic amine networks. Influence of curing cycles. Eur Polym J 45:1524–1534
10. Merdas I, Thominette F, Tcharkhtchi A, Verdu J (2002) Factors governing water absorption by composite matrices. Compos Sci Technol 62:487–492
11. Adamson MJ (1980) Thermal expansion and swelling of cured epoxy resin used in graphite/epoxy composite materials. J Mater Sci 15:1736–1745
12. Enns JB, Gilham JK (1983) Effect of the extent of cure on the modulus, glass transition, water absorption, and density of an amine-cured epoxy. Appl Polym Sci 28(9):2831–2846
13. Gupta VB, Drzal LT, Rich M (1985) The physical basis of moisture transport in a cured epoxy resin system. J Appl Polym Sci 30(11):4467–4693
14. Johncock P, Tudgey GF (1986) Some effects of structure, composition and cure on the water absorption and glass transition temperature of amine-cured epoxies. Brit Polym J 18(5): 292–302
15. Van Krevelen DW, Te Nijenhuis K (2009) Properties of polymers. Their correlation with chemical structure. Their numerical estimation and prediction from additive group contributions, 4th edn. Elsevier, Amsterdam
16. Small PA (1953) Some factors affecting the solubility of polymers. J Appl Chem 3(2):71–80
17. Van Krevelen DW (1965) Chemical structure and properties of coal. XXVIII. Coal constitution and solvent extraction. Fuel 44(4):229–242
18. Hoy KL (1970) New values of the solubility parameters from vapor pressure data. J Paint Techn 42(541):76–118
19. Fedors RF (1974) A method for estimating both the solubility parameters and molar volumes of liquids. Polym Eng Sci 14:147–154
20. Barrie JA (1968) Water in polymers. In: Crank J, Park GS (eds) Diffusion in polymers, Chap. 8, Academic Press, London, pp 259–314
21. Gaudichet-Maurin E, Thominette F, Verdu J (2008) Water sorption characteristics in moderately hydrophilic polymers, Part 1: Effect of polar groups concentration and temperature in water sorption in aromatic polysulfones. J Appl Polym Sci 109(5):3279–3285
22. Courvoisier E (2017) Analysis and kinetic modeling of the thermal ageing of PEI and PEEK matrices and its consequences on water absorption. PhD thesis, Arts et Métiers ParisTech, Paris, March 2017

23. Baschek G, Hartwig G, Zahradnik F (1999) Effect of water absorption in polymers at low and high temperatures. Polymer 40:3433–3441
24. Thominette F, Gaudichet-Maurin E, Verdu J (2008) Effect of structure on water diffusion in moderately hydrophilic polymers. Defect Diffus Forum 258(260):442–446
25. Carter HG, Kibler KG (1978) Langmuir-type model for anomalous moisture diffusion in composite resins. J Compos Mater 12(2):118–131
26. Cai LW, Weitsman Y (1994) Non-Fickian moisture diffusion in polymeric composites. J Compos Mater 28(2):130–154
27. Brethous R, Colin X, Fayolle B, Gervais (2016) Non-Fickian behavior of water absorption in an epoxy-amidoamine network. AIP Conf Proc 1736: paper no. 020070
28. Serpe G, Chaupart Verdu J (1997) Ageing of polyamide 11 in acid solutions. Polymer 38 (8):1911–1917
29. El Mazry C, Correc O, Colin X (2012) A new kinetic model for predicting polyamide 6-6 hydrolysis and its mechanical embrittlement. Polym Degrad Stab 97(6):1049–1059
30. Li L, Yu Y, Wu Q, Zhang G, Li S (2009) Effect of chemical structure on the water sorption of amine-cured epoxy resins. Corros Sci 51(12):3000–3006

Influence of Glass Fibre Sizing and Storage Conditions on Composite Properties

Luc Peters

Abstract This paper describes a major component of glass fibre reinforcements, the sizing, and its impact on mechanical properties of epoxy or polyester laminates. First the basics of the glass fibre manufacturing process and the types of silanes used for direct rovings are briefly described. The important feature of these coatings is that they can be affected by storage conditions. The study focuses on the effect of roving storage time, temperature and humidity as well as packaging type on the fibre/matrix interphase and on the inter-fibre properties of the composites. The conclusions show that not all fibre sizings offer the same storage stability and that warehousing and packaging details can have a significant impact on the final composite performance.

Keywords Glass fibre · Sizing · Storage · Ageing · Moisture

1 Introduction

3B is part of the Braj Binani Group which is a conglomerate with diversified interests in cement, zinc and glass fibre. 3B is a major actor in composite reinforcement solutions, with a special focus on thermoplastics, wind energy and performance composites [1]. The company has a long heritage of 45 years of expertise and produces more than 160,000 tonnes of glass fibre products per year.

Glass fibre production was developed in the 1930s and has been extensively documented in various textbooks and websites [e.g. 2–4]. The basic process involves melting silica sand, limestone, kaolin clay and other minerals to liquid form at temperatures above 1500 °C, then extruding through bushings to make

L. Peters (✉)
3B Fibreglass, Battice, Belgium
e-mail: luc.peters@3b-fibreglass.com

P. Davies and Y.D.S. Rajapakse (eds.), *Durability of Composites in a Marine Environment 2*, Solid Mechanics and Its Applications 245,
https://doi.org/10.1007/978-3-319-65145-3_2

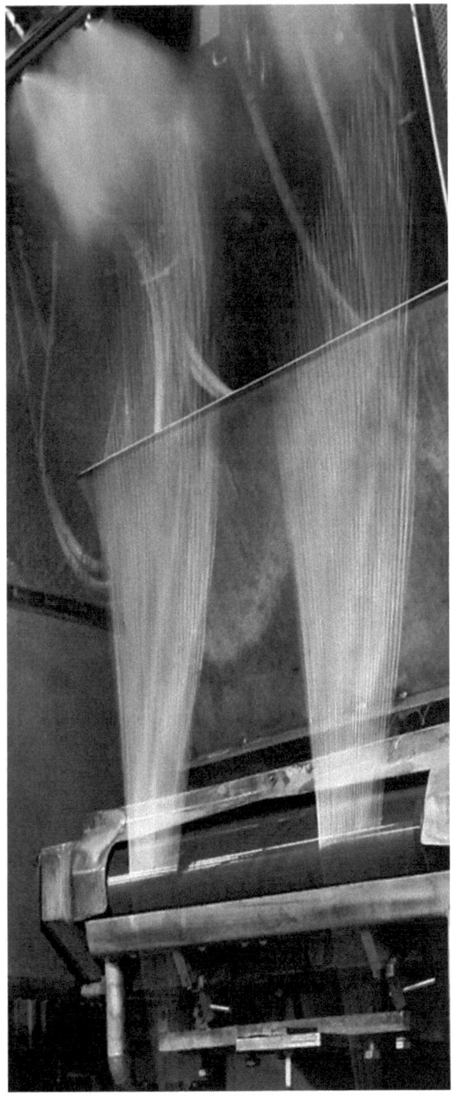

Fig. 1 Glass fiberizing and coating with sizing

fibres, Fig. 1. The most common fibre is E-glass, which is alumino-borosilicate glass with less than 1% by weight of alkali oxides. 3B produces two unique high performance and eco-responsible glass technologies: Advantex® glass, a corrosion resistant grade (E-CR) [5], and HiPer-tex™, a fibre with improved mechanical properties [6].

Advantex® is a registered trademark of Owens Corning used under license.

Figure 2 shows a schematic diagram of the process.

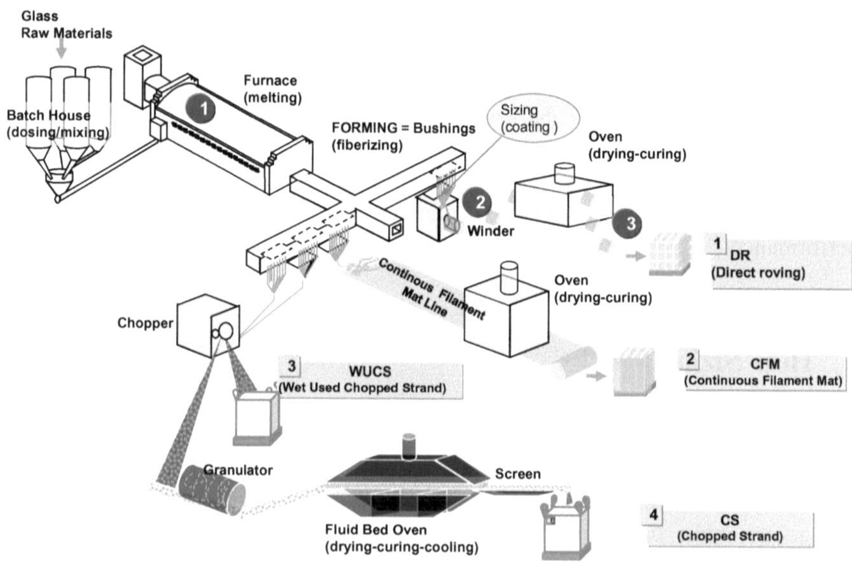

Fig. 2 Schematic figure showing glass fibre production

2 Sizing Composition

Size is a coating applied as a water-based formulation, it contains

- Coupling agent(s): Provides adhesion between glass surface and the resin (usually silanes),
- Film Former(s): Provides protection and strand integrity to the roving as well as compatibility with the resin (epoxy, polyester, vinylester, etc.),
- Lubricant(s): Provides lubrication and protects the filaments during processing,
- Other Additives: antistatic, emulsifier, anti-foaming, bactericide, etc.

Sizing is applied to the fibres as they are drawn down from the hot bushings, as a water-based emulsion with 2–10% of solid. Final solid content on the dry fibre can range from 0.3 to 1.5% depending on final product application and sizing type. Table 1 shows a typical formulation.

Table 1 Typical glass fibre sizing composition

Component	% of sizing
Silane(s)	5–15%
Film former	50–70%
Lubricant	10–30%
Additives	0–5%

The application of this sizing to the fibres is an incredibly complex operation, as Thomason has pointed out [7]. The fibre bundle is pulled at velocities up to 60 m/s, so sizing pick-up occurs in less than 0.5 ms. The sizing must then fully impregnate the bundle before it reaches the first winding sheave, i.e. in less than 0.2 s.

After coating, the fibres are wound onto bobbins, then dried and cured in an oven.

Optimising the sizing is a balancing act between cost, processing and performance.

3 How Sizing Works

The role of the glass sizing is to ensure good transfer of load applied on the composite from the polymer matrix to the glass. Glass fibre (GF) sizing is specific to the polymer matrix and the final application. Table 2 shows the difference a good sizing can make to an injection molded short fibre reinforced polyamide 66 specimen.

The most common sizing strategy is to use condensation of silanes onto the glass surface as shown in Fig. 3.

Different types of silane are used for different thermosetting matrix resins, such as γ-aminopropyltriethoxysilane for epoxies and γ-methacryloxypropyltrimethoxysilane for unsaturated polyester and vinylesters. The reactive groups from the silane (shown in green on the figures in the electronic version) will react with the resin or hardener reactive groups.

These are shown below, Fig. 4.

Table 2 Example of two glass fibre reinforced PA66 composites with different sizings

Injection molded PA66-50%wt GF (Chopped strands reinforcement)	Poor coating	Good coating
Tensile strength (MPa)	50	180
Notched Impact strength (kJ/m^2)	<5	18
Interface quality	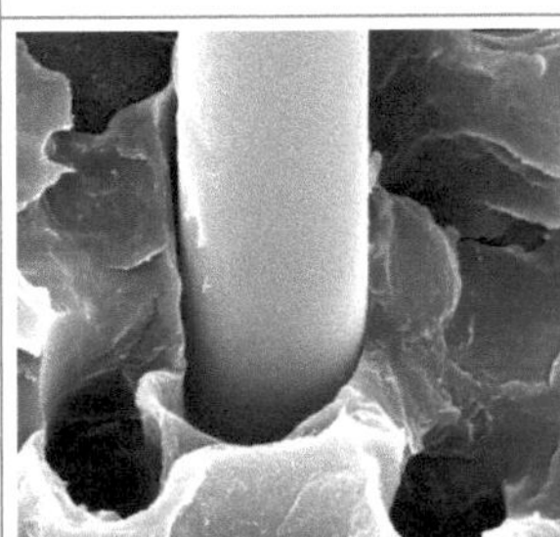	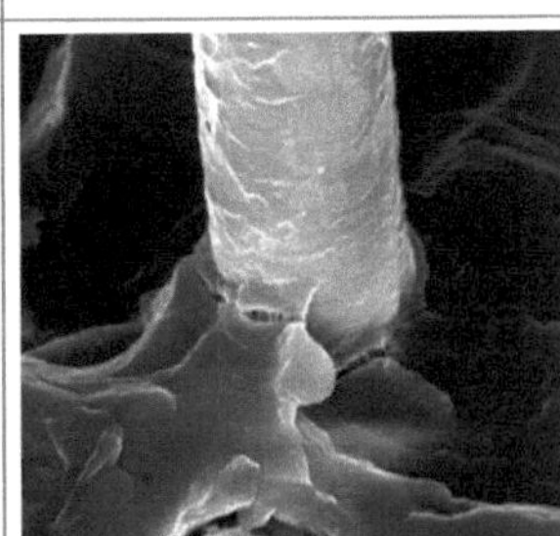

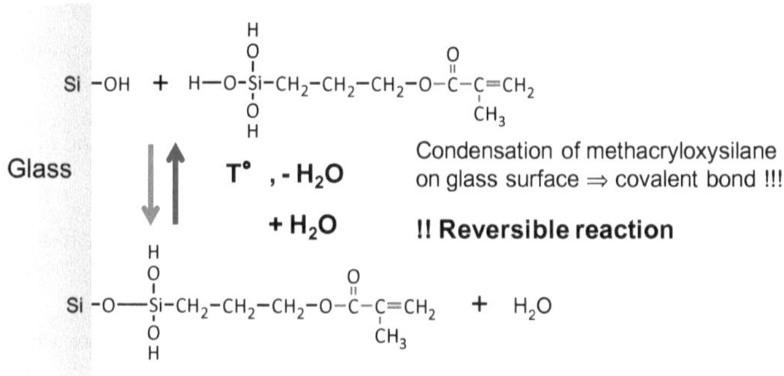

Fig. 3 Application of silane to glass surface

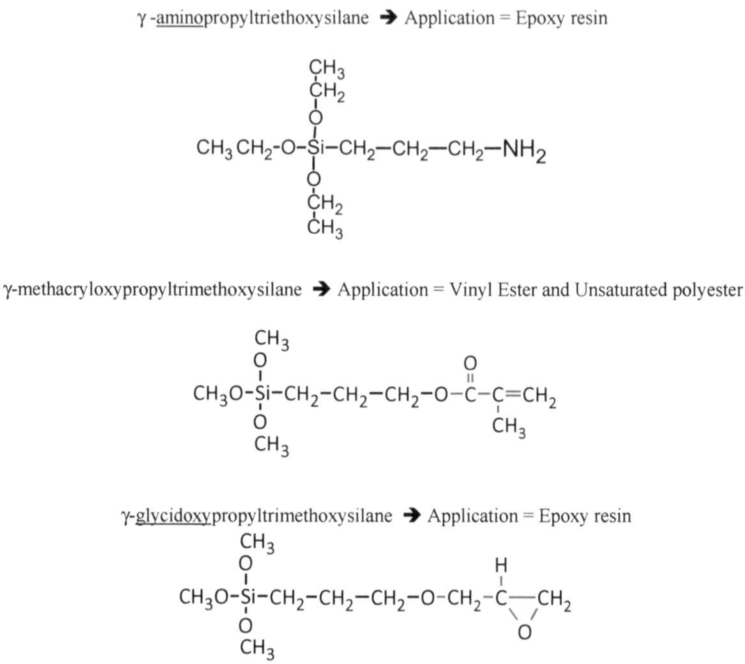

Fig. 4 Typical coupling agents for thermoset matrix composites

4 Long Term Stability of Sizing on Glass Fibres

Continuous fibres used for the reinforcement of thermoset resins show usually a decrease of adhesion properties with storage conditions and time (Figs. 5 and 6). This results in an important deterioration of transverse tensile strength and interlaminar shear strength as highlighted later in this article (Figs. 7 and 8).

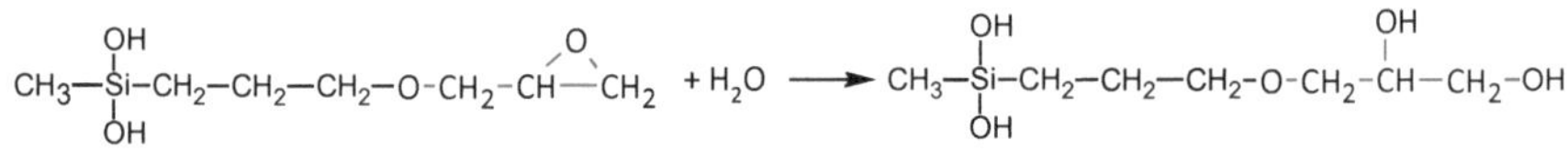

(a) Hydrolysis of Epoxysilane when exposed to moisture and heat

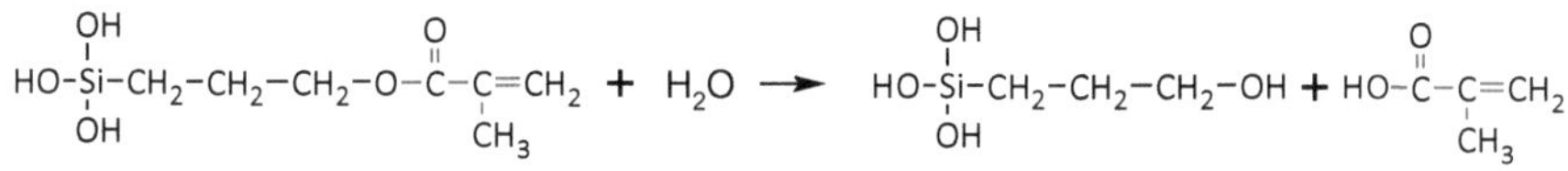

(b) Hydrolysis of Methacryloxysilane when exposed to moisture and heat

Fig. 5 Examples of sizing reactivity reduction due to hydrolysis of sizing ingredients when exposed to moisture and heat

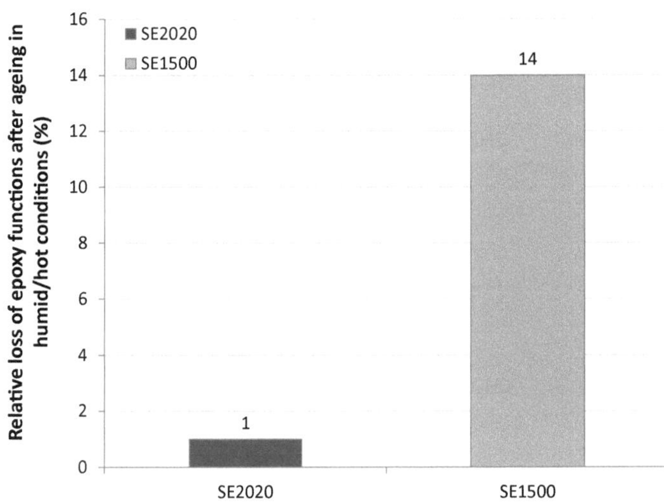

Fig. 6 Relative loss of epoxy functions (peak ref. = 912 cm^{-1}) after ageing of sizing films in humid and hot conditions (30 °C—80%RH) during 12 weeks

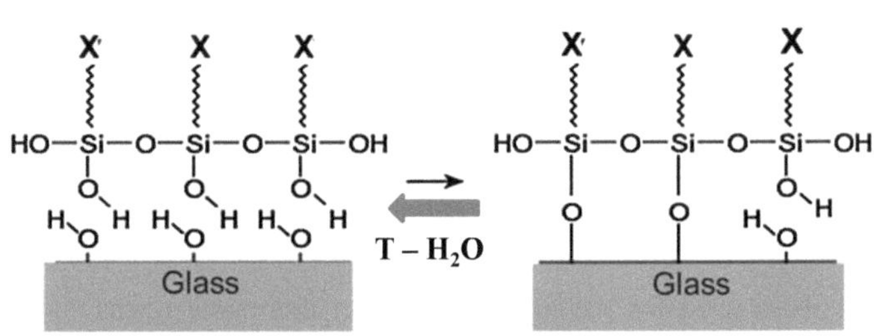

Fig. 7 Hydrolysis mechanism of Si–O–Si bonds at glass—sizing interface when exposed to moisture and heat

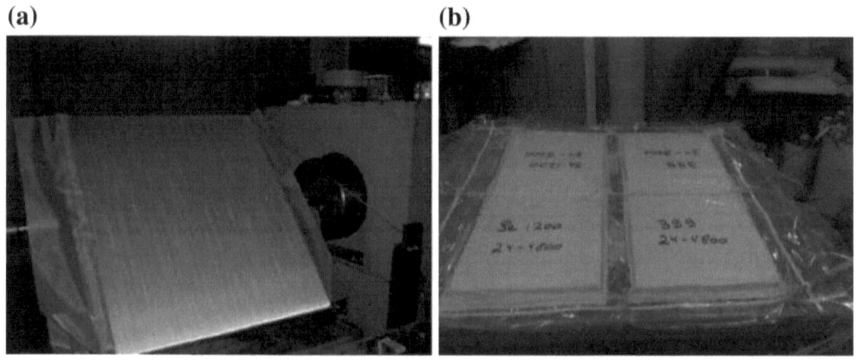

Fig. 8 Sample preparation. **a** reinforcement winding, **b** Infusion (4 products/infusion in most cases)

This decrease of adhesion properties is clear evidence that sizing chemistry on the glass is affected by the storage conditions. There are several mechanisms that could explain this drop of properties

(A) Hydrolysis of chemical functions when exposed to moisture and heat
(A.1) Sizing-matrix interface: reduction of sizing reactivity with the matrix.

The role of the sizing is to chemically bond the glass surface with the polymer matrix. This implies that some sizing ingredients will react with the resin during the processing of the composite. Nevertheless, the majority of the chemical functions used for that purpose are sensitive to moisture and heat. Those so-called hydrolysis reactions will result in a reduction of the sizing reactivity with the polymer matrix and consequently in a decrease of fibre-matrix adhesion. The chemicals functions that are the most impacted are the epoxy functions (epoxy silane and bisphenol-A epoxy resin usually used as film former) and the methacryloxy function (methacryloxysilane). The reaction schemes of epoxy and methacryloxysilane hydrolysis are presented in Fig. 5. Similar mechanism to epoxysilane takes place in the case of bisphenol-A epoxy resin.

The hydrolysis of the epoxy functions has been demonstrated by FTIR at the level of 3B-Fibreglass laboratory on different sizing films stored at 30 °C and 80% RH during 12 weeks. Figure 6 presents the relative loss of epoxy functions for films of SE2020 and SE1500 stored at 30 °C—80%RH compared to reference films stored at 24 °C—47%RH.

Based on those results, it is clear that the sensitivity to moisture and heat depends on the sizing composition. Indeed, the recently developed SE2020 sizing is clearly less affected by the epoxy hydrolysis reactions than an old generation sizing like SE1500.

(A.2) Glass-matrix interface: hydrolysis of Si–O–Si bonds

As presented in Fig. 3, the functionalized silanes react with the silanol bonds present at the surface of the glass. Unfortunately, this condensation reaction is reversible and the formed Si–O–Si bonds could be hydrolyzed when the fibres are exposed to moisture and heat. This results in a weakness at the level of the glass-sizing interface and consequently in a decrease of adhesion properties at the level of the composite. Figure 7 illustrates again the reaction scheme of reversible equilibrium hydrolysis-condensation of Si–O–Si bonds.

(B) Glass-matrix interphase: reduction of sizing solubility when exposed to heat in dry environment

Glass fibre sizings for thermoset applications usually contain a high concentration of epoxy functions. When exposed to heat in dry conditions, those epoxy functions could react with other sizing ingredients bearing, e.g. hydroxyl, amino or carboxylic acid functions. Those post-reactions will generate high molecular species that will reduce the sizing solubility with the liquid resin during composite manufacturing. This phenomenon can cause impregnation issues and defects that will negatively impact the mechanical properties.

In the case of sizing formulations containing methacryloxysilane or more generally unsaturations, heat will accelerate the oxidation of the double bonds and the creation of radicals. The presence of radicals could lead to cross-linking reactions that will again reduce sizing solubility.

On top of that, the two mechanisms described here before will have as consequence a reduction of the reactive species concentration on the fibres.

There is little published information on the long-term stability of glass fibre sizings in the scientific literature. While the importance of the interphase region between fibre and matrix is generally accepted the complexity of the chemistry in that zone has limited the number of studies. Ishida and Koenig [8] published some early results showing loss of silane coating after an 80 °C water treatment.

Plueddemann [9] discussed some of the factors that affect water resistance of coupling agents, and Ishida provided a review of the subject in 1984 [10].

Wang et al. [11] used mass spectroscopy to study the interphase in E-glass surfaces coated with silanes and exposed to different hot wet conditions. They revealed the presence of layers of different hydrolytic stability. In a second paper [12] using X-ray photon spectroscopy, they showed evidence of aluminium ion migration from the fibre surface into the silane layers, and these layers remained after hot water extraction.

Salmon and colleagues at ENSAM, Paris [13, 14] described a series of tests on two common coupling agents for amine-cured epoxy/glass fibre composites: APS (γ-aminopropyltriethoxysilane) with amine groups and GPS (glycidylpropylsilane), in which the coupling function is an epoxide group (glycidyl). These were taken as model compounds for part of the interphase and subjected to wet ageing under various conditions at temperatures from 0 to 100 °C. They clearly showed that the organosilane layer cannot act as a protective barrier to water ingress.

Given the lack of available results, and also the specific nature of the conditions to which glass fibre reinforcements can be subjected before and after manufacturing, 3B launched a detailed study to quantify how the stability of composites is affected by the storage and packaging conditions of the sized glass reinforcements.

5 Materials and Methods

Fibre reinforcements

The composites tested were based on the following fibre products:

Pallets of fibres from the 3B Birkeland plant (rovings have different production dates):

Reference sizing (SE1500) 17–2400, production date: 4/3/2013.

New sizing (SE2020) 17–2400, production date: 11/4/2012.

New Multi-Compatible sizing (W3030)17–2400, production date: 5/9/2013.

Note: 2020 sizing is epoxy specific, and 3030 sizing is multi-resin compatible (UP, VE, EP).

"SE" indicates that the sizing is applied on Advantex glass, "W" means it is applied on HiPer-tex glass.

Ageing conditions

Trip to/from Dubai for SE1500 and SE2020 from August to December.

1 week at −18 °C (to check Epoxy crystallisation).

2.5 months in oven at 50 °C in Laboratory (RH = 2–3%).

2.5 months oven 50 °C +2.5 months at 30 °C/80%RH (21.5 g H_2O/kg air).

2.5 months oven 50 °C +2.5 months at 30 °C/80%RH in two sealed PE bags.

Composite panel manufacture

After ageing, composite panels were manufactured from each batch of fibres for testing.

Glass fibres were aligned by dry winding of flat panels, Fig. 8a:

Four layers of 845 g/m^2 UD rovings for a reinforcement weight of 3380 g/m^2

Laminates of 2 mm thickness were produced by infusion with Hexion RIMR 135/137 epoxy resin at 35 °C under full vacuum, Fig. 8b, with a 4 h post-cure at 90 °C.

Glass contents were determined to be in the weight range of 72–75.5% in all cases.

The test methods used to evaluate the effect of sized fibre ageing on composite performance were

Tensile testing at 90 °C to the fibre direction, and

Short beam shear test according to ISO 14130.

These tests are known to be very sensitive to the fibre/matrix interface integrity.

6 Test Results

The environmental conditions during the return trip to Dubai were measured using a data logger, which indicated a temperature range from 8 to 42 °C and humidity from 40 to 80%.

Figure 9 shows the transverse tensile test results for all UD samples.

This plot provides a very large amount of information. It is clear that some panels have very low transverse tensile strengths, <10 MPa, this was due to transverse cracks already visible in the panels after manufacture. The improved sizing (2020 and 3030) show good strength retention even after 2 years, while the SE1500 properties (diamond symbols) tend to drop under all ageing conditions.

Figure 10 shows the interlaminar shear strength results.

The ordinate scale is reduced, and even in the worst case an ILSS value above 40 MPa is obtained, but the overall trends correspond very closely to those seen in transverse tension, with significant ageing effects for the SE1500 reference sizing.

It is also of interest to examine how such changes in interface performance might affect fatigue performance, as this is critical for the wind turbine industry. Figure 11 shows an example of test results. The green curve is an actual curve for a UD fabric with W2020 roving which was used for the certification of wind turbine blades. The red curve is an estimation of the possible drop in the case of sizing degradation; such drops have been seen with old sizings, such as SE1500 [15].

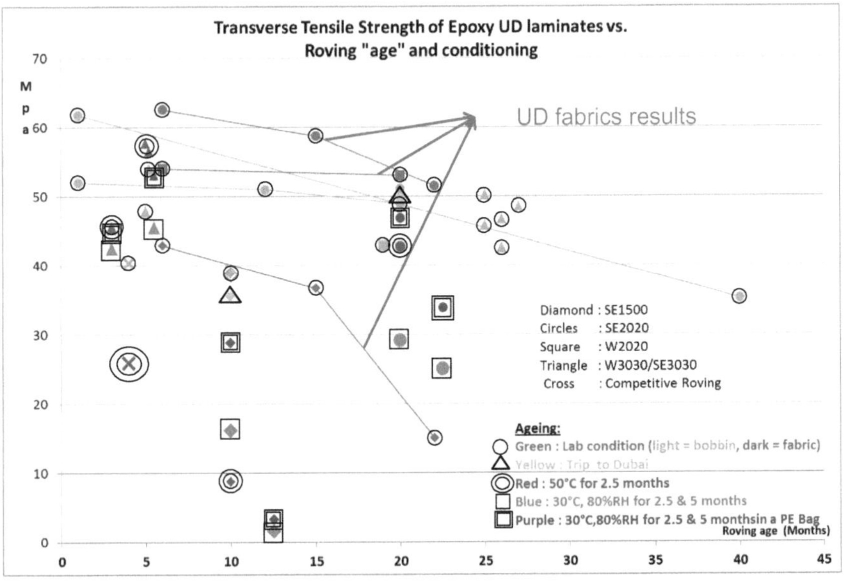

Fig. 9 Transverse tensile test results (Colours shown in electronic version)

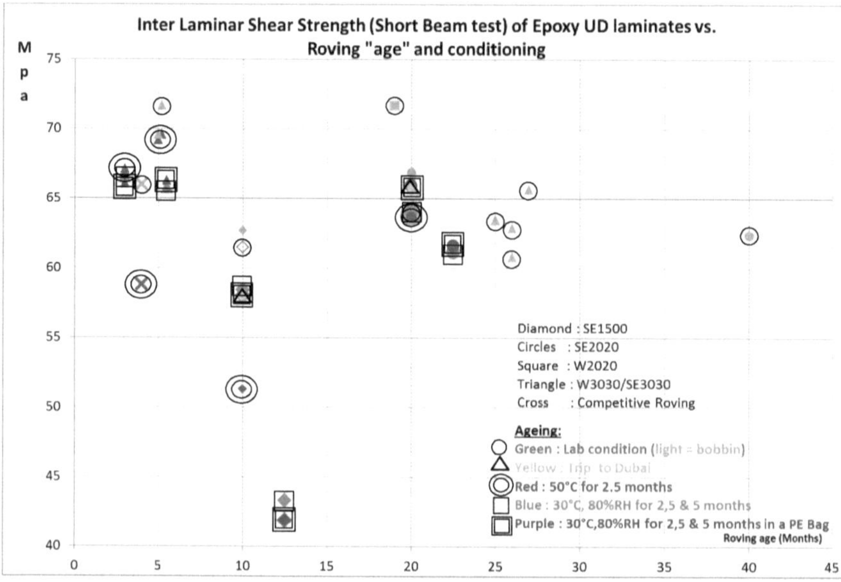

Fig. 10 Interlaminar shear strengths (Colours shown in electronic version)

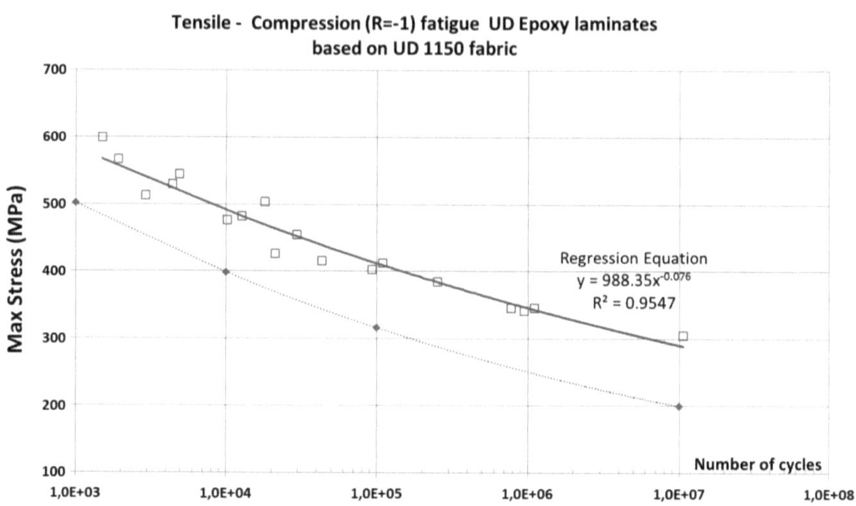

Fig. 11 Fatigue test results. New sizing test data (*green squares*) and degraded old sizing estimation (*red diamonds*)

7 Conclusions

This paper describes results from a study of the stability of sizings on glass fibres prior to manufacturing, and the consequences for composite interfacial properties.

As noted in service, some old reference sizings were confirmed to be very sensitive to storage conditions (especially temperature) and demonstrated significant reduction of interfacial properties over time.

The newly developed sizings demonstrate much reduced degradation over time in both rovings and fabric forms of Advantex and HiPer-tex glass fibres. These are much less sensitive to temperature, and provide significantly higher initial interfacial strength.

After 2 years of storage under normal Western European storage conditions, these new sizings still demonstrate excellent interfacial properties.

Degradation in hot and humid environments still occurs, most likely at the interface with the glass surface (hydrolysis of silane) but also at the level of the silane and film formers reactive groups. Tight packaging can help but water will still diffuse through eventually. Storage under controlled moisture levels can also help (i.e. <25 g H_20/kg or 35 °C/70%RH).

Newly developed sizings ensure better laminate property consistency and should be preferred for critical structural applications.

References

1. http://www.3b-fibreglass.com/
2. Vaughan DJ (1998) Fiberglass reinforcement. In: Peters ST (ed) Handbook of composites, Chap 7. Chapman and Hall Second edition
3. Jones FR (2001) Glass fibres. In: Hearle JWS (ed) High performance fibres, Chap 6. Woodhead
4. http://www.compositesworld.com/articles/the-making-of-glass-fiber
5. http://www.3b-fibreglass.com/3b-e-cr/performances-properties/
6. http://www.3b-fibreglass.com/hipertex/
7. Thomason JL, Adzima LJ (2001) Sizing up the interphase: an insider's guide to the science of sizing. Compos A 32:313–321
8. Ishida H, Koenig JL (1980) Effect of hydrolysis and drying on the silane bonds of a silane coupling agent deposited on E-glass fibers. J Polym Sci, Polym Phys Edn 233–237
9. Plueddemann EP (1981) Principles of interfacial coupling in fibre-reinforced plastics. Int J Adhes Adhes 1(6):305–310
10. Ishida H (1984) A review of recent progress in the studies of molecular and microstructure of coupling agents and their function in composites, coatings and adhesive joints. Polym Compos 5:101–123
11. Wang D, Jones FR, Denison P (1992) Surface analytical study of the interaction between γ-amino propyl triethoxysilane and E-glass surface. Part I. Time-of-flight secondary ion mass spectrometry. J Mat Sci 27:36–48
12. Wang D, Jones FR (1993) Surface analytical study of the interaction between γ-amino propyl triethoxysilane and E-glass surface. Part II. X-ray photoelectron spectroscopy. J Mat Sci 28:2481–2488

13. Salmon L, Thominette F, Pays MF, Verdu J (1997) Hydrolytic degradation of model networks simulating the interfacial layers in silane coupled epoxy/glass composites. Compos Sci Technol 57(8):1119–1127
14. Salmon L, Thominette F, Pays MF, Verdu J (1999) Hydrolytic aging of polysiloxane networks modeling the glass fiber epoxy-amine interphase. Polym Compos 20(6):715–724
15. Boisseau A, Davies P, Thiebaud F (2013) Fatigue behavior of glass fibre reinforced composites for ocean energy conversion systems. Appl Compos Mater 20(2):145–155

Water Uptake in Polymer Composites with Voids

Leif A. Carlsson and E. Du

Abstract Long-term durability assessment of polymer matrix composite materials exposed to humid air or liquid water requires a firm understanding of the mechanisms governing water uptake and the chemical interactions involved in degradation of the matrix and the fiber/matrix interface. The added mass of water will cause swelling of the matrix, which tends to degrade bonding between fiber and matrix. Voids provide space where water may accumulate, and if they are located at the fiber/matrix interface, the accumulated water may cause hydrolysis. Voids, furthermore, are geometrical irregularities that cause stress concentration, and reduction of strength. Fickian diffusion is considered valid to analyze the moisture uptake in polymers and void-free composites. For composites containing voids, however, deviations from Fickian diffusion are observed. Voids in the form of capillaries along the fibers define flow channels and promote 'wicking', which is a rapid water uptake mechanism. Experimental studies on water uptake in polymer matrix composites show a strong dependence of voids. A composite containing even a small void volume fraction will absorb moisture in excess of that contained in the matrix resin. The moisture uptake in a composite immersed in seawater has been modeled using a combination of diffusion of moisture in the matrix resin, combined with the capillary flow in the voids. The predictions qualitatively agree with experiments, but saturation times are off. More detailed characterization of the void structure and improved flow modeling are required to accurately predict the dynamics of water uptake in a composite.

Keywords Polymer matrix composite · Voids · Fiber–matrix interface · Water uptake · Fickian diffusion · Capillary flow

L.A. Carlsson (✉) · E. Du
Department of Ocean and Mechanical Engineering, Florida Atlantic University, Boca Raton, FL 33431, USA
e-mail: carlsson@fau.edu

P. Davies and Y.D.S. Rajapakse (eds.), *Durability of Composites in a Marine Environment 2*, Solid Mechanics and Its Applications 245,
https://doi.org/10.1007/978-3-319-65145-3_3

1 Introduction

Fiber-reinforced composite materials are used in multiple industrial applications including commercial, military, aerospace, naval, automotive, etc. Such materials offer marine applications significant advantages over metal alloys, such as reduced maintenance costs due to the high corrosion resistance of composite materials. However, the performance of fiber-reinforced composites with a polymer matrix is affected by several environmental factors such as elevated temperatures, ultraviolet rays, and immersion in seawater. Such degradation must be understood and considered for in design of composite structures.

Polymer matrixes are typically degraded by long-term exposure to humidity and water. The polymer matrix tends to absorb moisture which causes swelling (increase in volume) and reduction of modulus and strength of the material. Swelling of the resin due to absorption of water in the resin will relax residual compressive stresses at the fiber/matrix interface caused by cure shrinkage, and increase the possibility of debonding of the fiber/matrix interface. There are two main mechanisms for water transport in polymer matrix composites; diffusion through the matrix [1], and absorption through voids, micro cracks, and gaps between fiber [2–5]. Moisture transport in capillaries along the fiber/matrix interface is known as "wicking" [5, 6].

The objective of this chapter is to review moisture transport in composite materials reinforced with glass and carbon fibers. The Fickian diffusion model is typically adequate for modeling of moisture transport in the neat resins and has also been found applicable for modeling of moisture transport in polymer matrix composites with small void content [1]. For composites containing voids, it is demonstrated that they strongly influence the water uptake mechanisms.

2 Voids in Polymer Composite Materials

Voids are empty spaces in a material. In most cases, voids are undesired because they tend to reduce strength and provide spaces where absorbed water may accumulate. As will be discussed, voids in fiber composites are typically extended and run along the fibers in a composite providing flow channels that may contribute to more rapid water transport and increased accumulation of water. A primary source for void formation in polymer matrix composites includes the entrapment of air and volatile components evolving from the resin during liquid processing of the composite. If the viscosity of the liquid resin is high, flow around the fibers and filling of the spaces between the fibers become difficult. As a consequence, the resin may not adequately wet the fibers forming unbonded regions at the fiber/matrix interface [7]. Furthermore, it must be recognized that it is difficult to join two highly dissimilar materials, such as inorganic fibers to an organic polymer matrix. Rather than two surfaces being chemically bonded, a joint consisting of a transitional material between bulk fiber and bulk matrix called "interphase" is formed [8].

As an example, vinylester resins do not bond well to carbon fibers, and in addition, such resins undergo substantial shrinkage upon cure which has the consequence that the resin in high fiber volume fraction composites may pull away from the fibers and create voids at the fiber/matrix interface [9, 10]. Voids may form also in the matrix as isolated cavities, and may also be found between poorly bonded plies in laminated composites. Figure 1 shows optical and SEM micrographs of voids in composites.

Voids in a composite may be detected and analyzed using techniques such as the density method, optical inspection of cross section cuts, and ultrasonic C-scan [11]. The ASTM method [12] measures void content based on the mass per unit volume of the sample. Voids have no mass but contribute to increase the volume of the sample. The void-free density of a composite, however, is difficult to measure and is commonly calculated from known values of the fiber and resin densities and the weight fraction of fibers. To isolate the fibers in the sample and determine their weight, the matrix may be burned-off or dissolved using acid digestion methods [13], but the acid may degrade the fibers and remove the mass, which will lead to inaccurate results. For small void contents, the accuracy of this method becomes questionable and it is not uncommon to obtain negative values of the void content [7] which of course is impossible.

Microscopy requires cutting and inspection of an interior cross section of a sample (Fig. 1). This method allows analysis of void size, shape and content but is limited by its destructive character and uncertainties associated with the transformation of 2D surface analysis to 3D volumetric results. Many cross sections need to be inspected in order to obtain representative results for the entire sample.

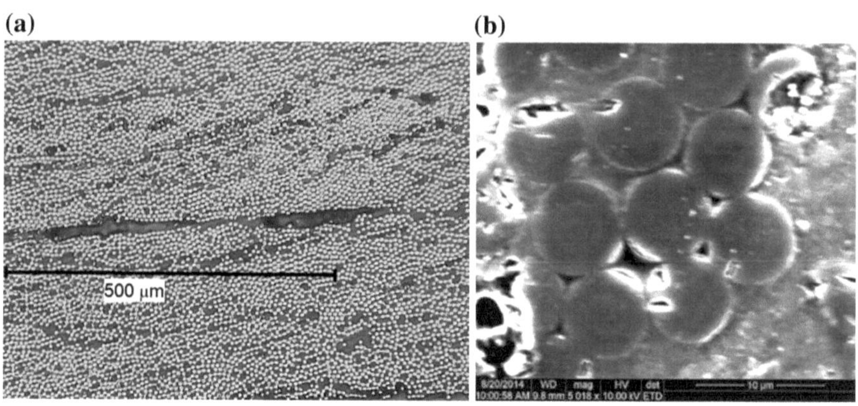

Fig. 1 Voids in a polymer matrix composite: **a** optical micrograph of cross section showing isolated voids (courtesy of Dr. Gillespie, CCM), **b** SEM micrograph of cross section of unidirectional carbon/vinylester composite cut perpendicular to fiber direction showing interfiber voids [28]

Little et al. [11] used X-ray computed tomography (μCT) to characterize the 3D size and distribution of voids in carbon–fiber composite laminates. This is a novel nondestructive technique suitable for detailed NDE of composites [14]. μCT imaging works by recording the attenuation of an X-ray beam passing through a sample with voids [15]. A 2D cross-sectional image is mathematically reconstructed from attenuation measurements made at different angles. From results determined from 2D slices of known separation, it is possible to construct a 3D model allowing for precise determination of void size, shape, and location. A scanning resolution of 7 μm was used in the Little et al. analysis of voids in a composite laminate [11]. Although it is likely that a large number of voids are smaller than 7 μm, such small voids are considered to have a minor influence on the quality of the laminates. Their analysis revealed the presence of large elongated voids, while the smaller voids were scattered throughout the entire volume. Comparison between the location and orientation of the elongated voids and the fiber angles in the laminates shows that the voids align with the fibers. This is consistent with the 2D view provided by the SEM micrograph of a cross section cut perpendicular to the fiber axis of a unidirectional composite, see Fig. 1b, where voids are observed between closely packed fibers. Apparently, such voids form channels between the fibers.

3 Moisture Transport Mechanisms

When polymers and polymer matrix composites are immersed in water they will absorb moisture until a saturation state is asymptotically reached. The absorption of moisture is generally very slow, and thick specimens may require several months of immersion before they are saturated. It is very difficult to monitor the exact moisture distribution inside polymer and composite panels. A simple measure of the overall moisture uptake is therefore commonly used, such as the percent weight gain, "moisture content", defined by

$$M\ (t) = \frac{W_t - W_0}{W_0} \tag{1}$$

where W_t is the weight of the specimen after immersion time t, and W_0 is the initial dry weight of the specimen [1, 5].

3.1 *Moisture Transport by Diffusion*

Diffusion is the process of moving water molecules from a region of high concentration to a low concentration region inside the material, and this process is driven by a concentration gradient. Experimental studies on polymers and polymer matrix composites indicate that in many cases the Fickian diffusion model [16] accurately predicts moisture uptake in neat resins and composites with low void content.

For a large thin plate immersed in water, moisture transport occurs essentially through-thickness and may be considered as one-dimensional (1D) [1, 16]. Crank [16] conducted a mass balance analysis of a volume element of the plate absorbing moisture through the thickness to derive Fick's second law,

$$\frac{\partial c}{\partial t} = D\frac{\partial^2 c}{\partial z^2} \tag{2}$$

where c is the concentration of moisture, $c = c(z, t)$ in which t is the time of immersion, z is the thickness coordinate of the thin plate, and D is the diffusivity. The diffusivity quantifies the flow of water molecules across a specified surface, and the rate at which moisture is absorbed. The diffusivity may be considered as a constant for a given temperature, type of fluid, and material. The diffusivity depends on the composition and temperature of the fluid the material is immersed in and the chemical composition of the material.

For a large and thin plate of thickness, h, where $h \ll w, l$ (w and l are the width and length of the panel), immersed in water, the Fourier solution of Fick's law (Eq. 2) shows that at short immersion times the moisture content varies linearly with the square root of the immersion time, t, and approaches the saturation moisture content, M_m, asymptotically.

For thicker plates of orthotropic material, such as a fiber-reinforced composite, Fig. 2, all exposed surfaces will contribute to the overall moisture uptake. The diffusivities,D_1, D_2, and D_3 represent diffusion through surfaces normal to the 1, 2 and 3 principal material directions. Unidirectional composites are often approximated as transversely isotropic materials; $D_2 = D_3 = D_T$ where T represents the direction transverse to the fiber. Furthermore, $D_1 = D_L$ where L represents the longitudinal direction (along the fiber). For this case, the effective diffusivity becomes [1, 17],

$$\sqrt{D} = \left(\frac{w + h}{w}\right)\sqrt{D_T} + \left(\frac{h}{l}\right)\sqrt{D_L} \tag{3}$$

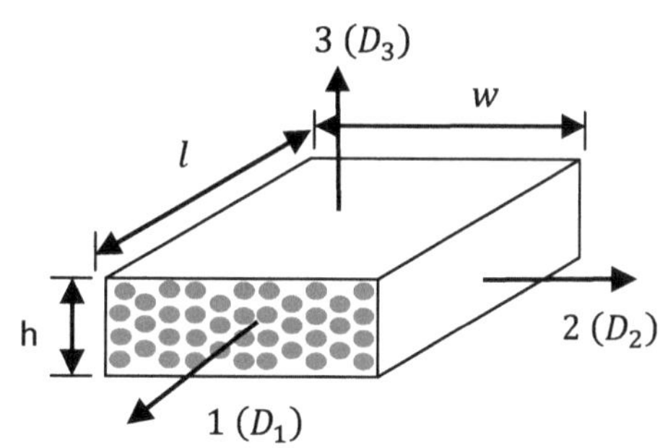

Fig. 2 Thin composite plate

For a perfectly bonded (void-free) transversely isotropic composite with fibers that are inert to moisture, D_T and D_L may be estimated from the resin diffusivity D_r using the following micromechanics expressions [1],

$$D_T = D_r\left(1 - 2\sqrt{\frac{v_f}{\pi}}\right) \tag{4a}$$

$$D_L = D_r(1 - v_f) \tag{4b}$$

where v_f is the fiber volume fraction.

The Fickian model has been applied also to damaged composites. A common damage mode observed during moderate loading of a multidirectional composite laminate is multiple matrix cracks in the 90° plies. These cracks are generally not detrimental because the strong 0° plies remain intact. However, if a cracked laminate is exposed to humid air or water, the matrix cracks provide channels for moisture transport to the interior of the composite. Lundgren and Gudmundson [2, 3] analyzed cracked cross-ply (0°/90°) composite laminates exposed to humid air. The model was used to determine the importance of 90° matrix cracks for the overall moisture uptake. They developed a two-phase moisture diffusion model where one phase is the homogenized periodically cracked 90° layers and the other phase the uncracked 0° plies. Each phase is assumed to be described by the Fickian diffusion model. For a laminate containing matrix cracks, the overall moisture uptake deviates from the Fickian model. If the number of cracked plies is large, the matrix cracks will start to dominate the moisture uptake process.

3.2 *Capillary Moisture Transport*

Extended voids at the fiber/matrix interface may define flow channels for moisture transport. Moisture transport in trees for example occurs in capillaries and is known as "wicking" [5, 6]. Kosuri and Weitsman [6] were the first to suggest capillary motion of water in composites. Cross-ply laminates (laminates consisting of only 0° and 90° plies) were slowly loaded along the 0° direction to a point where the (weaker) 90° plies cracked at multiple locations. Cracks that run along the fibers and span the entire thickness of the 90° ply (or 90° ply group) propagated until arrested at the 90°/0° interface. The width (opening) of these crack is about 1 μm thus acting as capillaries able of transporting water.

The cracked specimens were turned 90° so that the cracks assumed a vertical orientation. The bottom edge of each specimen was brought in contact with slightly acidic seawater to allow vertical capillary flow through the microcracks. A litmus paper was placed on the top horizontal edge to allow tracking of the time required for the capillary climb through the specimen. A range of climb heights from 1 to

4 cm was examined. Figure 3 shows results for the climb time versus distance for the specimens examined.

It is noted that the climb time versus distance results are approximately linear, in agreement with the capillary flow model discussed later in this section. Furthermore, the rate of capillary transport for this case is about 4 mm/min. Thus, the time required for the capillary flow front to travel 1 mm is about 0.25 min. This rate is about 5–6 orders of magnitude faster than the transport due to diffusion through the matrix [5].

Weitsman [4] examined an antisymmetric cross-ply carbon/epoxy composite laminate of lay-up $[0/90/0_4/90_4/0/90]_T$ subject to cyclic moisture exposure between 0 and 95% RH at 66 °C with a cycle period of 9 days. Damage in the form of fiber/matrix debonds was observed during repeated humidity cycling. A continuum damage model was developed for a unidirectional composite that absorbs moisture from a humid environment. The effects of damage on moisture transport were treated using thermodynamics analysis of an open system, and relations between absorbed moisture content, stress and damage were derived. The experiments on moisture absorbed in the absence and presence of external load verified several aspects of the damage model. Quantitative comparison between the model and experiments was not possible because the fiber/matrix debonding process was not considered in the model.

A composite material with elongated cracks and voids may be considered a porous material. Fluid transport in porous materials has been described and analyzed by Dullien [18]. Specifically, if such a material is fully or partially immersed in water, modeling of moisture transport may utilize the theory of capillary flow. The analysis of transport of water through a capillary is in its most simple form

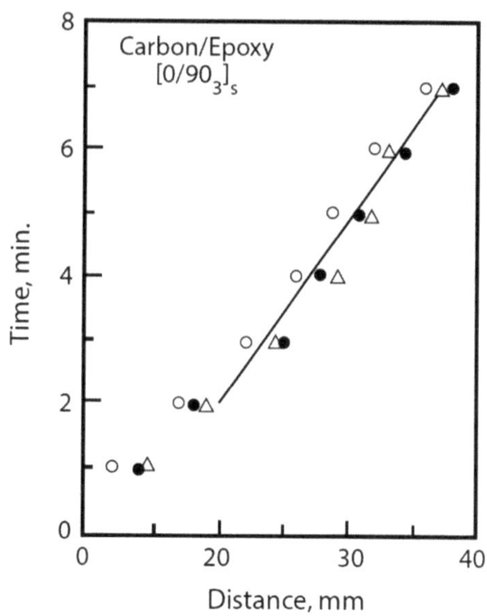

Fig. 3 Time for capillary flow of seawater through microcracks in $[0/90_3]_s$ carbon/epoxy laminates. Data from Kosuri and Weitsman [6]

based on the classical case of pressure driven laminar uncompressible fluid flow through a pipe, see Fig. 4. The average velocity, V_{avg}, is given by [19],

$$V_{avg} = -\frac{d^2}{32\mu}\left(\frac{dp}{dx}\right) \tag{5}$$

where d is the internal diameter of the pipe, $\frac{dp}{dx}$ is the pressure drop along the pipe, and μ is the viscosity of the fluid.

The volumetric flow rate, $Q\left(\frac{m^3}{s}\right)$, through the pipe with a cross-sectional area, A_c, is given by

$$Q = V_{avg}A_c = \frac{d^2\Delta p}{32\mu l}\left(\frac{\pi}{4}d^2\right) \tag{6}$$

This is identified as the classical Hagen–Poiseuille equation [18, 19],

$$Q = \frac{\pi d^4 \Delta p}{128\mu l} \tag{7}$$

For a cross section of porous material containing n capillaries, each of diameter d, the flow rate becomes

$$Q = n\frac{\pi d^4 \Delta p}{128\mu l}. \tag{8}$$

3.3 *Moisture Uptake of Syntactic Foam*

Syntactic foams consist of hollow glass microspheres embedded in a polymer resin. The size of the microspheres is statistically distributed over a range of diameters from about 10–100 μm [20]. The wall thickness of the glass shell is about 1–2 μm. The volume fraction of microspheres in a syntactic polymer foam is typically 0.4–

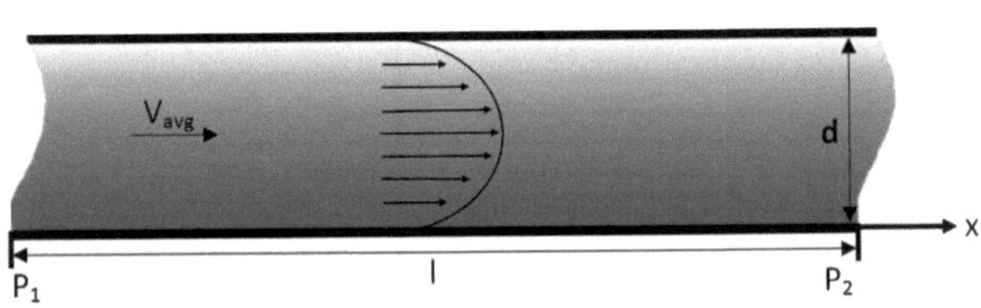

Fig. 4 Illustration of laminar pipe flow

0.6. Such materials have found deep sea offshore underwater application by their low thermal conductivity and low density.

Lefebvre et al. [21] and Sauvant et al. [22] examined water uptake in syntactic foam materials immersed in water over a range of temperatures and pressures. Specimens with and without glass microspheres were prepared using a high T_g (175 °C) epoxy resin matrix. The microspheres were sodium–borosilicate glass with a mean diameter of 34 μm and density of 0.35 g/cm^3. The volume fraction of the glass spheres was 0.55. Samples of dimensions 50 × 50 × 2 (mm) were immersed in water at temperatures from 20–120 °C over 18 months, and the moisture content was monitored by regularly weighing the samples.

It was found that the neat resin samples absorbed 1.2–1.7% moisture at saturation, depending on the temperature. The particle filled specimens reached much higher moisture contents, close to 40% although they were not fully saturated after 18 months. The results suggest that water did not only hydrate the matrix, but hydrated the glass/polymer interface, allowing diffusion of water into the glass shell, and filling the void inside the microsphere.

A model called IFP after Institut Français du Petrole was proposed by Lefebvre et al. [21]. The model is not based on first principles and includes three adjustable empirical parametric that allow construction of a moisture content versus time curve for microsphere filled polymers immersed in water. This model utilizes a representative volume element (unit cell) of the material containing a cube of polymer containing a single glass sphere in the center as schematically illustrated in Fig. 5.

The model includes three basic mechanisms, (i) Fickian diffusion through the matrix, (ii) matrix/particle shell hydration governed by an equilibrium reaction between the water absorbed in the matrix and water at the matrix/particle shell interface, and (iii) filling of the empty space (void) inside the particle by water.

The model was calibrated by adjusting the three parameters to a limited set of experimental data. The model predictions of moisture uptake versus time were then compared to experimental results for specimens immersed in water at 100 °C. Particle volume fractions of 0.30, 0.45 and 0.55 were examined. The model predictions were also compared to results for specimens with a constant particle volume fraction (0.55) for a range of temperatures, and good agreement was observed. Experimental and model results are also presented for syntactic foams immersed in

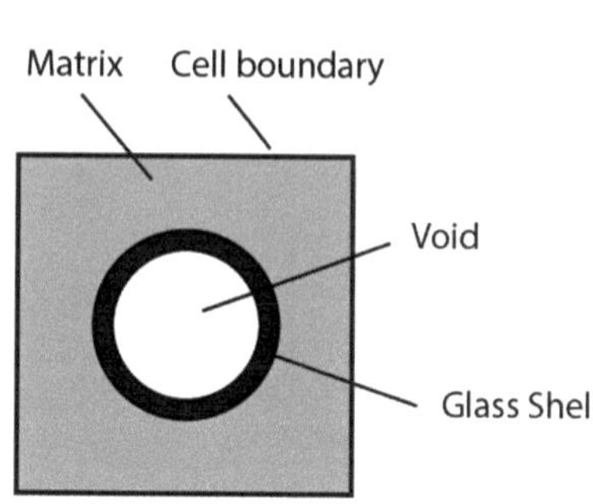

Fig. 5 Representative element of a syntactic foam [21]

seawater at various temperatures and pressures by Lefebvre et al. [21]. The results indicate that increased pressure will increase the rate and amount of water uptake.

4 Analysis of Excess Water Absorption in Voidy Composites

The moisture content is defined by (Eq. 1),

$$M(t) = \frac{m_w(t)}{m_{dry}} \tag{9}$$

where $m_w(t)$ is the mass of absorbed water after immersion time t, and m_{dry} is the initial dry mass of the material.

For a fiber composite, the dry mass is given by

$$m_c = \rho_f V_f + \rho_m V_m \tag{10}$$

where ρ_f and ρ_m are the densities of fiber and resin (dry) and V_f and V_m are the volumes of fiber and resin in the composite. Hence, the moisture content of a composite can be expressed as

$$M_c = \frac{m_w(t)}{\rho_f V_f + \rho_m V_m} \tag{11}$$

If we assume the fibers are inert, the resin in the composite is chemically the same as the neat resin, and at saturation, the voids fill up with water, the mass of absorbed water in the composite, at saturation, becomes,

$$m_{ws} = M_{ms} m_m + \rho_w V_v \tag{12}$$

where M_{ms}, m_m, ρ_w and V_v are the saturation moisture content of the matrix, the dry mass of the matrix, the density of water, and the total volume of voids in the composite. Substitution of Eq. (12) in (11) gives the saturation moisture content of the composite,

$$M_{cs} = \frac{M_{ms} m_m + \rho_w V_v}{\rho_f V_f + \rho_m V_m} \tag{13}$$

The void volume can be expressed as,

$$V_v = v_v V_c \tag{14}$$

where v_v is the void content and V_c is the volume of the composite,

$$V_c = V_f + V_m + V_v \tag{15}$$

The overall density of the composite is given by,

$$\rho_c = \frac{m_c}{V_c} \tag{16}$$

where m_c is the mass of the composite. Specifically, at dry conditions, m_c is given by

$$m_c = m_m + m_f \tag{17}$$

Substitution of Eq. (10) into (16) yields the density of the dry composite,

$$\rho_c = \frac{\rho_f V_f + \rho_m V_m}{V_c} \tag{18}$$

This equation simplifies to

$$\rho_c = \rho_f v_f + \rho_m v_m \tag{19}$$

where v_f and v_m are the volume fractions of fiber and matrix, respectively.

Substitution in Eq. (13) gives the saturation moisture content of the composite,

$$M_{cs} = \frac{M_{ms} m_m + \rho_w v_v V_c}{\rho_c V_c} \tag{20}$$

This equation simplifies to

$$M_{cs} = \frac{M_{ms} m_m}{\rho_c V_c} + \frac{\rho_w v_v}{\rho_c} \tag{21}$$

The weight fraction of (dry) matrix is given by,

$$w_m = \frac{m_m}{m_c} \tag{22}$$

Notice that for the dry composite,

$$w_f + w_m = 1.0 \tag{23}$$

where w_f is the weight fraction of fiber. This equation is valid also for dry composites with voids because empty voids have no mass.

Substitution of Eq. (22) into (21) yields,

$$M_{cs} = w_m M_{ms} + \frac{\rho_w}{\rho_c} v_v \tag{24}$$

If the saturation moisture content of the composite is normalized with the resin weight fraction, we get

$$\frac{M_{cs}}{w_m} = M_{ms} + \frac{\rho_w}{\rho_c w_m} v_v \tag{25}$$

Hence, this normalization should provide a direct measure of the void content. This will be discussed further later in this section.

The influence of voids on moisture transport in several glass/epoxy composites has been studied by Thomason [7]. Glass/epoxy composites with a fiber volume fraction of about 0.65 were examined. Specimens in the form of rods were prepared by a pultrusion technique. A range of void contents was obtained by varying the cure pressure [23]. The void content was measured using the ASTM D2734 density method [12]. Composite and matrix densities were determined from the weight difference of samples in air and ethanol. The density of the glass fibers was measured after burning off the coating at 500 °C. The void contents ranged from 0.74 to 9.51% [7, 23]. The specimens were exposed to a relative humidity (RH) of 100% for approximately 1 year. Moisture content plotted versus square root of time (according to the Fickian model) revealed non-Fickian absorption in the composites with voids. Table 1 summarizes saturation moisture contents as measured by Thomason [7] for composite B over a range of void contents from 0.8–3.5%. The moisture contents normalized with the weight fraction resin ($w_m = 0.2$) is also listed.

Table 1 Saturation moisture contents for glass/epoxy composite (B) at various void contents [7], and normalized moisture content (M_{cs}/w_m). Saturation moisture content of neat epoxy (M_{ms}) is 2%

Void content (%)	M_{cs} (%)	M_{cs}/w_m (%)
0.80	0.44	2.2
1.76	0.80	4.0
2.75	1.30	6.5
3.51	1.54	7.7

Based on the normalized moisture contents, it is obvious that the composite absorbs substantially more moisture than the matrix resin (2%), and the excess moisture content increases with increasing void content.

For the composite "B" (w_m = 0.2), Eq. (24) becomes,

$$M_{cs} = 0.004 + 0.474\mathrm{v_v} \tag{26}$$

Moisture content predictions from this equation and experimentally measured values for composite B are plotted versus void content in Fig. 6.

The hypothesis that the voids are filled up with water is supported by the experimental results, although it is noted that the experimental results fall below the prediction. The difference, less than 0.5%, may be due to the exposure of the composite to humid air (100% RH) as opposed to immersion in water and the void filling may require more time.

Further results on the relation between voids and moisture absorption have been presented by Figliolini and Carlsson [24] who examined moisture uptake in neat vinylester 510A resin and a large set of composite specimens consisting of T700 carbon fibers with two sizings (F and G) in a vinylester 510A resin (denoted C(F)/VE510A and C(G)/VE510A) immersed in seawater at 40 °C. The saturation moisture content of neat VE510A resin immersed in seawater at 40 °C is 0.344%. The fiber volume fractions for the C(F)/VE510A and C(G)/VE510A were 0.66 and 0.63.

To model the saturation moisture content of the composite with voids, equation (23) may be expressed as,

$$M_{cs} = \frac{\rho_m v_m}{\rho_c} M_{ms} + M_{vs} \tag{27}$$

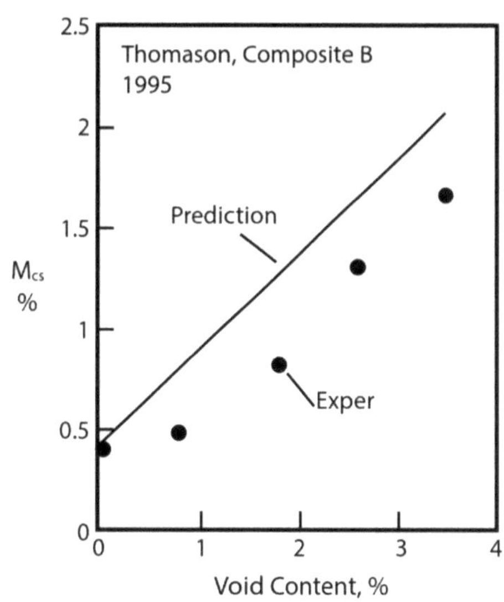

Fig. 6 Saturation moisture content of glass/epoxy composite ("B") versus void content. The straight line represents Eq. (26) and the points are measured data [7]

Table 2 Saturation moisture contents and excess moisture contents in carbon/vinylester composites [24]. M_{cs} and M_{vs} are the measured moisture content of the composite and the excess moisture content due to voids

	C(F)/VE510A	C(G)/VE510A
M_{cs} (%)	0.470	0.565
M_{vs} (%)	0.398	0.486

where M_{vs} is the excess moisture content due to the filling of the voids. Analysis of the moisture uptake data for the two types of composites using Eq. (27) provides the results listed in Table 2.

It is noted that moisture uptake for the carbon/vinylester composites is greatly dominated by voids (84–86%). It is also noted that the type of fiber sizing influences the moisture uptake. The composite with the G sized fibers absorbed more water than the one with F sized fibers.

Although interface voids appears to be interconnected throughout the composite, there may be some isolated voids. For isolated voids to be filled with water, individual water molecules would first have to diffuse through the composite matrix and then fill the voids. This process is expected to be very slow compared to wicking.

4.1 Influence of Pressure and Temperature on Moisture Absorption

The influence of hydrostatic pressure on composite samples immersed in water has been examined by several authors. Tucker et al. [25, 26] found that the influence of excessive pressures on water absorption is small in composites with low void content. However, for composites containing voids, recent results presented by Humeau [27] show a strong influence of pressure, where increased pressure leads to faster water uptake and higher moisture contents.

Figure 7 shows moisture uptake results for specimens immersed in tap water at 20 and 60 °C, with and without an added pressure above atmospheric pressure.

At 20 °C, the moisture uptake appears Fickian, with a saturation of about 0.7% after 2 years. At 60 °C and no added pressure, the moisture uptake increased continuously without an indication of saturation during 2 years of immersion. The specimens kept at 60 °C under 10 MPa pressure absorbed moisture more rapidly and reached higher saturation moisture contents (about 1.6%). The strong influence of pressure on the moisture uptake is attributed to voids in the filament wound glass/epoxy composite (void content of several percents). As discussed earlier, voids have a strong effect on the moisture uptake, Fig. 6. The wicking transport mechanism discussed earlier indicated that pressure is an important driving force, Eq. (8). For a glass/epoxy composite with very low void content produced from

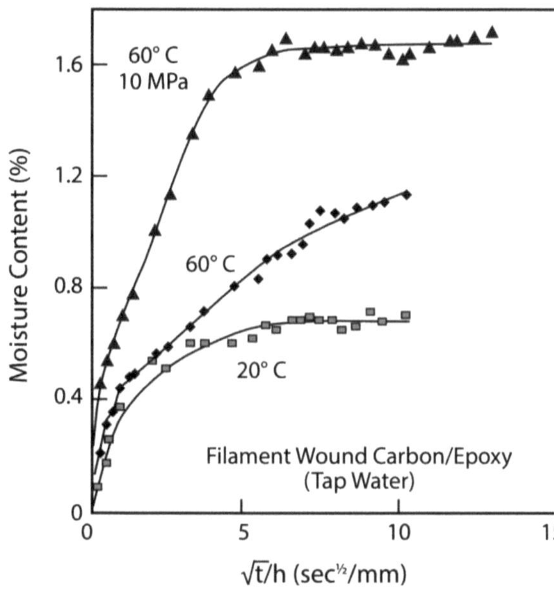

Fig. 7 Influences of pressure and temperature on moisture uptake of carbon/epoxy filament wound cylinders immersed in tap water [27]

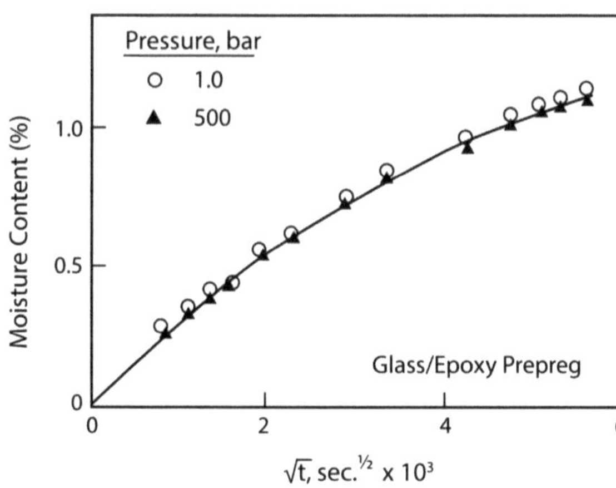

Fig. 8 Influence of hydrostatic pressure on moisture uptake of glass/epoxy composite made from prepreg [27]

prepreg, the influence of pressure on the moisture uptake is negligible [27], see Fig. 8.

The results in Fig. 7 show that temperature of the water has a strong effect on the rate of moisture uptake. Diffusion of moisture is a thermally activated process with the rate governed by an Arrhenius type of equation [1],

$$D = D_o e^{-A/RT} \tag{28}$$

where D_o is the reference diffusivity, A is the activation energy, R is the gas constant, and T is the temperature in °K.

4.2 Influence of Processing on Moisture Uptake

As discussed in the previous sections voids have a strong influence on the moisture uptake of composites. Voids are introduced during liquid resin processing of the composite as a result of air entrapment, inadequate impregnation of the fibers, or resin shrinkage during cure. As an example, the moisture uptake in two highly similar resin and composite panels, viz. neat VE510 resin and T700 carbon/VE510 panels prepared at University of Tennessee (UT) and Florida Atlantic University (FAU) are compared. All composite panels were prepared using vacuum assisted resin transfer molding (VARTM). The panels were immersed in seawater at 40 °C until saturation.

Figure 9 shows moisture uptake versus time for the panels. The UT neat resin panels absorbed much more water than the FAU neat resin panel. The difference may be due to different curing agents, and cure conditions. The FAU composite panels absorbed much more water than those made at UT. The high moisture content in the FAU composite panel is attributed to voids, see Fig. 1b. The exact mechanism for the void formation remains unknown, but must be attributed to resin/catalyst chemicals, mixing ratios, temperature, and vacuum that influence resin viscosity and flow during processing. It is also likely that Florida and Tennessee have different compositions of the sea water. The moisture content of the UT neat resin panels is greater than that of the UT composite panels which is consistent with a low void content.

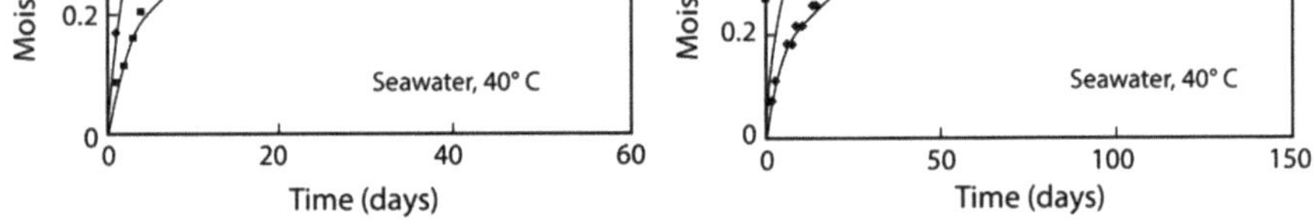

Fig. 9 Moisture uptake curve for neat VE510 and carbon/VE510 panels immersed in seawater at 40 °C. **a** University of Tennessee results, **b** Florida Atlantic University results

5 Modeling of Moisture Transport in Unidirectional Voidy Composite

5.1 Materials and Interface

Glass fibers are amorphous, inert materials that do not absorb water. Polymer matrix materials, such as vinylester and epoxy, are nonpolar thermosets that can absorb water by diffusion governed by Fick's law. Voids in composite materials have different sizes and geometries, caused by different mechanisms. Isolated voids in the matrix materials may be caused by air entrapment or by vaporization of volatile components during the manufacturing process. If the resin does not wet the fiber perfectly, voids may occur at the fiber/matrix interface. If the interface voids are connected, they may be considered as capillaries that run along the fiber axis in the composite material [28]. These voids in the composite materials provide additional pathways for water absorption, which explains the additional moisture uptake observed experimentally in carbon/vinylester composites [29, 30] as well as in the glass/epoxy composites [7].

During the manufacturing process, surface treatment and coating (sizing) of fibers is conducted to improve the fiber/matrix interfacial bonding, through primarily three mechanisms, including chemical adhesion, physical adhesion and mechanical adhesion [23]. For instance, glass fibers can be treated with an organosilane coupling agent, such as R-SiX3, which works as a chemical bridge [31] and a major component of the interphase formed between the fiber and matrix [32].

Utilization of coupling agents also improves the physical adhesion through a good wetting between the fiber and matrix. When the composite materials are exposed to water or moisture during service, the sizing chemistry and corresponding surface and interfacial energy may affect the kinetics of water transport at the interface. Experimental results in Table 2 showed that the type of fiber sizing influences both bonding and excess moisture uptake in T700 carbon fiber/vinylester (510A) composites with two separate sizings (F and G) [24].

Capability of surface wetting can be evaluated by contact angle measured at the air–liquid–solid interface. A contact angle below 90° represents high wettability, while a contact angle greater than 90° represents low wettability. Contact angle between neat resin and liquids can be measured directly by dispensing droplets on the resin surface. Neat resins are mostly hydrophobic. The contact angle of water on neat vinylester is about 91.8° [33].

Contact angle of liquids on single fibers of known diameter, d can be measured using the Wilhelmy balance method, in which a single fiber is oriented perpendicular to the liquid–air interface with a known surface tension, γ_{lv}, and the force, F exerted on it is measured. The contact angle is then calculated from the Wilhelmy equation,

$$F = \pi d \gamma_{lv} cos\theta \tag{29}$$

The contact angle of liquid resin decreases from 55° for native E-glass fibers to about 25° after a silane coating for thermoplastics is added [34]. The contact angle of water of those silane coated fibers ranges from 70° to 76° depending on the sizing chemistry [35]. As the values are less than 90°, the treated solid surface is hydrophilic and prefers to be covered by water as opposed to gas.

5.2 Langmuir-Type Sorption Model

The classical Fickian diffusion model has been commonly used to describe the diffusion in polymeric systems [1, 36]. For voidy composite materials, the interphace diffusion model is limited to predict the high diffusivity observed experimentally. A theoretical analysis by Piggott and Woo [37] indicated that a void path model with a diffusivity of 15 times that of the polymer was more suited to explain the experimental data for carbon/epoxy and glass/epoxy composites. Recently, López et al. [38] proposed a Langmuir-type model to describe the sorption mechanism in voidy materials. This model considers two separate parts of water sorption in the porous polymers, including a mobile fraction diffusing in the matrix and a bound fraction filling the voids, described by

$$Y_w = \frac{m_w}{m_0} = Y_w^{mat} + Y_w^{cav} \tag{30}$$

Two extreme conditions are considered for above equation. In the dry state, both water mass fractions are zero. In the wet equilibrium state, both parts are saturated with water, marked as $Y_{w(\infty)}^{mat}$ and $Y_{w(\infty)}^{cav}$. The water transport behavior through the matrix with presence of voids is then described by

$$\frac{\partial Y_w^{mat}}{\partial t} = -Ddiv\nabla Y_w^{mat} - \alpha \frac{\partial Y_w^{cav}}{\partial t} \tag{31}$$

where D is the classical Fickian diffusion coefficient of water in the matrix polymer, α is the water transfer coefficient between the diffusion in the matrix and filling in the voids. The second term on the right side of above equation describes the water trapped in the voids. To further model the kinetics of water filling in the voids, characteristic time, τ is introduced to derive a phenomenological expression of $\partial Y_w^{cav}/\partial t$,

$$\frac{\partial Y_w^{cav}}{\partial t} = -\frac{1}{\tau}\left(Y_w^{cav} - Y_{w(\infty)}^{cav}\frac{Y_w^{mat}}{Y_{w(\infty)}^{mat}}\right) \tag{32}$$

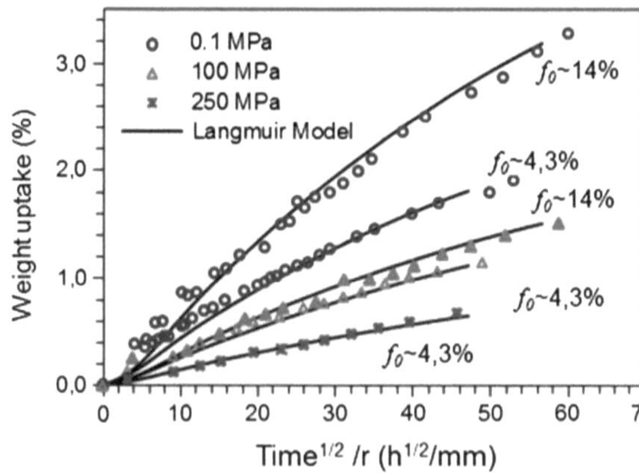

Fig. 10 Langmuir-type sorption modeling of water uptake in porous semicrystalline fluoropolymer samples [38]

Water sorption in porous semicrystalline fluoropolymer samples was experimented by gravimetric measurements at 40 °C and at varied pressure levels, including atmospheric pressure and hydrostatic pressures of 100 and 250 MPa. Matching the theoretical curves to the experimental measurements gives values of parameters α and τ. With fitting parameters $\tau = 100\,h$ and $\alpha = 0.2$ and 0.7 for two samples with initial porosity levels $f_0 = 4.3\%$ and 14%, respectively (Fig. 10).

5.3 *Capillary Flow Model*

A different model to describe water absorption in the voids of composite materials is capillary flow using Darcy's law (Eq. 8), where capillary flow is assumed in the voids along with each fiber at the fiber/matrix interface. Fichera and Carlsson [28] utilized the capillary flow model to explain seawater absorption in the voids of unidirectional carbon/vinylester (C/VE) 510A panels. Parameters for seawater absorption model predictions in unidirectional C/VE 510A panels were determined experimentally, including saturation moisture content M_m, void content V_v, saturation time as well as the dimensions of the capillaries (Table 3). Specifically, for the diffusion model, the effective diffusivity, D of the composite was calculated (Eq. 3) from transverse diffusivity, $D_T = 0.258 \times 10^{-6}$ mm^2/s and longitudinal diffusivity, $D_L = 1.00 \times 10^{-6}$ mm^2/s. As the panel size increases, the effective diffusivity decreases due to the higher value of D_L than D_T. For the capillary model, effective diameter, d and number of the capillaries was calculated based on several assumptions: the total volume of the capillaries equals to the void volume; capillary flow takes a path along with the fiber axis; moisture saturation occurs when both the matrix resin reaches saturation by diffusion and capillaries are filled by water.

Superposition of the capillary flow in the voids at the fiber/matrix interface and the classical Fickian diffusion-based water absorption in the polymer matrix gives a complete description of water absorption in the voidy composite materials

Table 3 Parameters for combined diffusion and capillary flow model predictions in C/VE 510A composite panels [28]

Parameter		Panel size (cm)		
		2 × 2	5 × 5	10 × 10
Saturation moisture content	M_m (%)	0.587	0.43	0.395
Void content	V_v (%)	1.00	0.54	0.46
Saturation time	t (day)	8	36	60
Effective diffusivity	D (10^{-7} mm^2/s)	6.15	4.05	3.3
Capillary number	n (10^6)	1.16	3.38	6.74
Capillary diameter	d (μm)	0.86	0.67	0.64

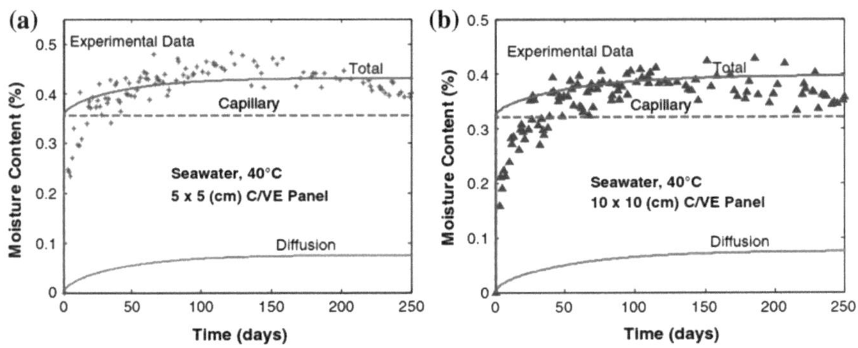

Fig. 11 Comparison of theoretical prediction of moisture content with the experimental measurements for **a** a 5 cm × 5 cm C/VE panel and **b** a 10 cm × 10 cm C/VE panel [28]

(Fig. 11). Both the diffusion and capillary flow models predict an increase in the saturation time with the panel size, agreeing with the experimental measurements. When the composite panels are exposed to seawater, water wicking starts immediately due to capillary flow in the voids of the exposed surfaces, followed by slow diffusion in the matrix material. However, a limitation of this model lies in the saturation time predicted by the capillary flow model, which is much shorter than that of the excess moisture determined from experiments. This is due to the assumption that the capillaries are uniformly distributed without any isolated voids and capillary flow runs along with the fiber axis at a constant speed.

The void content, determined from the excess water absorption decreases with the size of the unidirectional C/VE 510A panels. The saturation time for the excess water absorption, determined by subtracting the total water uptake with the diffusion fraction in the matrix, increases with the panel size. These observations suggest the presence of impedance to the capillary flow and storage in the voids at the fiber/matrix interface, likely from isolated voids or from the resistance of the trapped air when capillary flow advances from both openings of a capillary.

In that case, the smaller panels are likely to have more interconnected capillaries so that the saturation of the excess moisture beyond diffusion in the vinylester matrix is faster, as observed experimentally.

5.4 Future Work: Improved Capillary Model

Both Langmuir-type sorption model and capillary flow model share a common basis in describing the water diffusion in voidy composite materials, which divides the water absorption into two separate fractions: one moisture fraction modeled with classical Fickian diffusion and the other moisture fraction modeled with phenomenological expression of water sorption in voids (Langmuir-type model) or flow in open-ended capillaries at fiber/matrix interface (capillary flow model). However, both models are not well suited to accurately predict the kinetics of water filling in the voids for a number of reasons: (i) voids in a composite material may consist of both isolated voids and interconnective voids; (ii) location of the voids in the polymeric matrix or at the fiber/matrix interface determines the surface properties of voids and thus the interfacial properties for liquid flow within the void areas; (iii) air trapped within the interconnected voids would act against the capillary flow but eventually escape from the voids by slow perfusion.

We propose an improved capillary model to describe the kinetics of excess water absorption beyond the Fickian diffusion in a polymer matrix, which considers the effects of void size, location and surface properties. When a composite panel is immersed, water can enter the capillaries from both openings at the exposed surfaces (Fig. 12). Voids in the polymer matrix are considered as hydrophobic capillaries that repel water from entering in the channels, as the neat vinylester resin has contact angle of water greater than 90°. Voids at the fiber/matrix interface are hydrophilic capillaries as fibers treated with silane agents have contact angle of water less than 90°, so that water can enter immediately due to surface tension. Pressure of the air trapped inside the voids increases gradually with water filling. As the pressure difference across the water/air interface changes with the water front, as

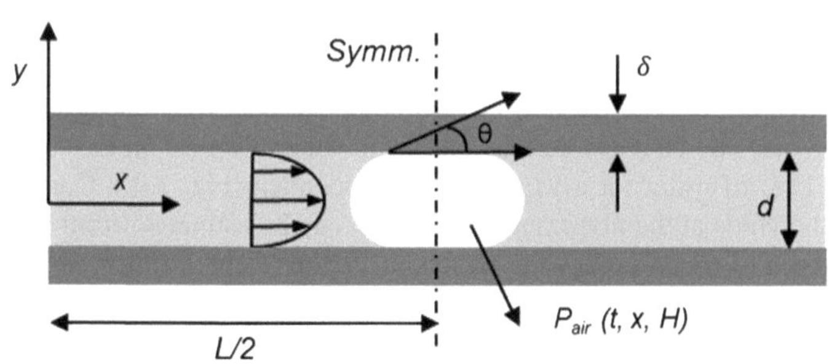

Fig. 12 Capillary flow in a void with both openings exposed to water

a result of a combination of viscosity, fluid inertia, and dynamic pressure, the filling rate is no longer a constant.

Assuming the flow in a single capillary is fully developed Poiseuille flow, the force balance on the capillary may be expressed as,

$$(m+\Delta m)\ddot{x} = 2\pi R\gamma cos\theta + \pi r^2 \cdot \rho g H - \pi r^2 \cdot P_{air} - \pi r^2 \cdot \frac{1}{2}\rho\dot{x}^2 - 8\pi\mu x\dot{x} \qquad (33)$$

where $m = \pi R^2 x\rho$, Δm is the added mass, contact angle θ of seawater in the capillary, $r = d/2$ is the radius of the capillary and H is the underwater depth of specimen. We further assume that the pressure of trapped air obeys Boyle's law,

$$P_{air} = P_0 \cdot V_0/\left(V_0 - 2\pi R^2 x\right) \qquad (34)$$

where P_0 is the atmospheric pressure, V_0 is the initial volume of trapped air in the capillary. The trapped air can slowly escape from the capillary by diffusion. The rate of volume loss of air can be calculated using Fick's law,

$$-\frac{dV_i}{dt} = \frac{SG_i\Delta p_i}{\delta} \qquad (35)$$

where subscript i represents the gas component in the air, S is the surface area for permeation of the trapped air, G_i is the permeability of the matrix polymer to each gas component, Δp_i is the difference in the partial pressures of the gas component between inside and outside of the capillary, δ is the thickness of matrix polymer film where gas permeation occurs. As indicated by the experimental study by Thomason [7], the void content is a dominant factor of water absorption in terms of the initial absorption rate and also in the final equilibrium level. The proposed capillary model could provide a quantitative analysis of the kinetics of water absorption in voidy composite materials.

The parameters for the capillary flow model can be determined experimentally, including size and location of the equivalent capillaries from μ-CT imaging [11] and surface properties from contact angle measurement (Fig. 13). The void content determined from μ-CT scan (Fig. 13a) can be validated by the experimental measurement, in that the amount of water absorbed by the voids within the composite panels can be calculated by subtracting the total water content with the moisture accumulated through the diffusion mechanism. The mass of absorbed water can then be converted to void content in the composite panels. Contact angle of water for matrix voids can be measured directly from water droplets on the surface of neat resin, e.g. 111° of water for a VE510A resin specimen (Fig. 13b). Contact angle of water for the voids at the fiber/matrix interface, can be estimated from water droplet on the surface of diced composite panels at various depths followed by polishing

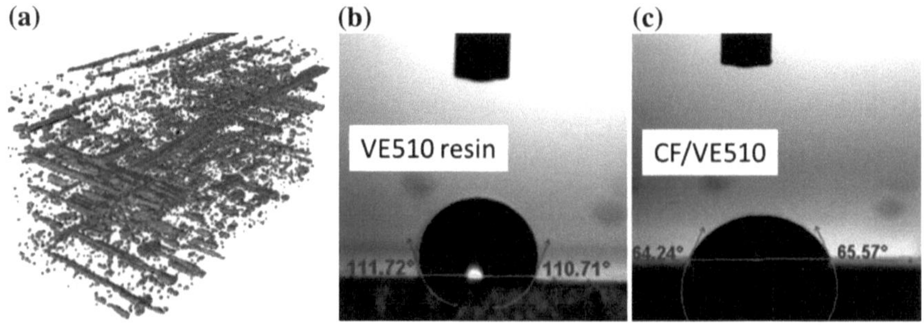

Fig. 13 Characterization of void content and surface properties: **a** μ-CT image showing structure of extended voids in a unidirectional composite [11]; **b** and **c** contact angle of water droplet on the surfaces of neat resin and composite materials

and cleaning processes, e.g. 65° of water for a carbon/VE510A specimen (Fig. 13c). For the interfiber voids, contact angle can be determined from the Wilhelmy balance method.

Acknowledgements Much of this work was supported by an ONR grant N00014-05-1-0341. Thanks are due to the ONR program officer, Dr. Yapa Rajapakse. Part of this material is based upon work supported by the National Science Foundation under Grant No. 1562062. Thanks go to Christopher Bellido for his help in compiling this chapter, and to Shawn Pennell for graphics support. The research by Maryann Fichera provided much insight into the role of capillary voids for moisture uptake.

References

1. Springer GS (1981) Environmental effects on composite materials. Technomic Publishing Co., Westport, CT
2. Lundgren JE, Gudmundson P (1998) A model for moisture absorption in cross-ply composite laminates with matrix cracks. J Compos Mater 32(24):2226–2253
3. Lundgren JE, Gudmundson P (1999) Moisture absorption in glass-fibre/epoxy laminates with transverse matrix cracks. Compos Sci Technol 59(13):1983–1991. doi:10.1016/S0266-3538(99)00055-X
4. Weitsman Y (1987) Coupled damage and moisture-transport in fiber-reinforced, polymeric composites. Int J Solids Struct 23(7):1003–1025. doi:10.1016/0020-7683(87)90093-X
5. Weitsman Y (2012) Fluid effects in polymers and polymeric composites. Mechanical engineering series. Springer, New York
6. Kosuri R, Weitsman Y (1995) Sorption processes and immersed fatigue response of gr/ep composites in sea water. In: Proceedings of the ICCM-10, Whistler, British Columbia, Canada, vol 4, pp 177–184
7. Thomason J (1995) The interface region in glass fibre-reinforced epoxy resin composites: 2. Water absorption, voids and the interface. Composites 26(7):477–485
8. Drzal LT (2001) Interfaces and interphases. Composites, ASM Handbook, Material Park, pp 169–179
9. Drzal L (2010) Engineering the carbon fiber-vinylester matrix interface for naval structural systems. ONR solid mechanics program review report, pp 23–32

10. Xu L, Mase T, Drzal L (2003) Influence of cure volume shrinkage of the matrix resin on the adhesion between carbon fiber and vinyl ester resin. Proceedings of ICCM-14, San Diego, 2003
11. Little JE, Yuan XW, Jones MI (2012) Characterisation of voids in fibre reinforced composite materials. Ndt&E Int 46:122–127
12. ASTM D2734, Standard test methods for void content of reinforced plastics (2009) ASTM International, West Conshohocken
13. Carlsson LA, Adams DF, Pipes RB (2014) Experimental characterization of advanced composite materials. CRC Press, Boca Raton
14. Wright P, Fu X, Sinclair I, Spearing SM (2008) Ultra high resolution computed tomography of damage in notched carbon fiber-epoxy composites. J Compos Mater 42(19):1993–2002. doi:10.1177/0021998308092211
15. Michael G (2001) X-ray computed tomography. Phys Educ 36(6):442–451
16. Crank J (1975) The mathematics of diffusion, 2nd edn. Clarendon Press, Oxford
17. Pomies F (1992) Degradation of composite materials in a marine environment. Master thesis, Florida Atlantic University
18. Dullien FAL (2012) Porous media: fluid transport and pore structure. Academic Press, New York
19. Cimbala JM (2006) Fluid mechanics: fundamentals and applications, vol 1. Tata McGraw-Hill Education, New York
20. Shutov FA (1986) Syntactic polymer foams. Adv Polym Sci 73–4:63–123
21. Lefebvre X, Sauvant-Moynot V, Choqueuse D, Chauchot P (2009) Durability of syntactic foams for deep offshore insulation: modelling of water uptake under representative ageing conditions in order to predict the evolution of buoyancy and thermal conductivity. Oil Gas Sci Technol 64(2):165–178. doi:10.2516/ogst/2008053
22. Sauvant-Moynot V, Gimenez N, Sautereau H (2006) Hydrolytic ageing of syntactic foams for thermal insulation in deep water: degradation mechanisms and water uptake model. J Mater Sci 41(13):4047–4054
23. Thomason JL (1995b) The interface region in glass-fiber-reinforced epoxy-resin composites. 1. Sample preparation, void content and interfacial strength. Composites 26(7):467–475. doi:10.1016/0010-4361(95)96804-F
24. Figliolini AM, Carlsson LA (2014) Mechanical properties of carbon fiber/vinylester composites exposed to marine environments. Polym Compos 35(8):1559–1569. doi:10.1002/pc.22809
25. Tucker WC, Brown R (1989) Moisture absorption of graphite polymer composites under 2000 feet of seawater. J Compos Mater 23(8):787–797. doi:10.1177/002199838902300802
26. Tucker WC, Lee SB, Rockett T (1993) The effects of pressure on water transport in polymers. J Compos Mater 27(8):756–763. doi:10.1177/002199839302700801
27. Humeau C, Davies P, Jacquemin F (2016) Moisture diffusion under hydrostatic pressure in composites. Mater Des 96:90–98
28. Fichera M, Carlsson LA (2016) Moisture transport in unidirectional carbon/vinylester panels with imperfect fiber/matrix interface. J Compos Mater 50(6):751–760. doi:10.1177/0021998315580834
29. Karbhari VM, Zhang S (2003) E-glass/vinylester composites in aqueous environments—I: experimental results. Appl Compos Mater 10(1):19–48
30. Ramirez FA, Carlsson LA, Acha BA (2008) Evaluation of water degradation of vinylester and epoxy matrix composites by single fiber and composite tests. J Mater Sci 43(15):5230–5242. doi:10.1007/s10853-008-2766-z
31. Chawla KK (2012) Composite materials: science and engineering. Materials research and engineering. Springer, New York
32. Jones FR (2007) The chemical aspects of fiber surfaces and composite interfaces and interphases, and their influence on the mechanical behavior of interfaces. In: The 28th Riso international symposium on materials science, Riso Nat'l Lab, Roskilde, Denmark, 2007

33. Alateyah AI, Dhakal HN, Zhang ZY (2013) Contact angle measurement of the vinyl ester matrix nanocomposites based on layered silicate. Int J Chem Nucl Metall Mater Eng 7(12): 614–619
34. Broutman L, Krock R, Plueddemann E (1974) Composite materials. Interfaces in polymer matrix composites, vol 6. Academic Press, New York
35. Van de Velde K, Kiekens P (2000) Wettability and surface analysis of glass fibres. Indian J Fibre Text 25(1):8–13
36. Shen C-H, Springer GS (1976) Moisture absorption and desorption of composite materials. J Compos Mater 10(1):2–20
37. Piggott MR, Woo M (1988) Water absorption of resins and composites: IV. Water transport in fiber reinforced plastics. J Compos Technol Res 10(1):20–24
38. López CC, Lefebvre X, Brusselle-Dupend N, Klopffer M-H, Cangémi L, Castagnet S, Grandidier J-C (2016) Effect of porosity and hydrostatic pressure on water absorption in a semicrystalline fluoropolymer. J Mater Sci 51(8):3750–3761

Durability of US Naval Composites and Sandwich Structures: Science Framework Considering Multiscale Response in Harsh Sea Environment

Dayakar Penumadu

Abstract Deformation behavior of carbon fiber-reinforced polymer composites involves mechanisms at multiple length scales with a strong dependency on the carbon fiber type and quality, fiber sizing, choice of polymer matrix, form of reinforcement, and manufacturing process. Naval structures typically employ sandwich materials including low-density foam core materials. Given the large amount of variables associated with materials and processing methods, a systematic science based framework is needed for predictive capabilities to model static, fatigue, and fracture behavior and its dependency on seawater and temperature. Using materials of high interest to US Navy, such a framework is proposed considering T700 carbon fibers specifically sized for vinylester resin system having brominated additives for improved fire performance. Textile composites manufactured utilizing VARTM process with suitable post-cure were utilized throughout the experimental campaign. The sandwich lay-up for this study consists of a closed cell polymeric (PVC) foam core placed between thin carbon fiber-reinforced vinylester facings. A fundamental study to evaluate the environmental degradation associated with long-term exposure on mechanical properties of carbon fiber-(Toray T700 3K tow-based stitched fabric) reinforced vinylester (CF/VE) facings at different orientations and various specimen sizes was implemented. The tension–tension fatigue effects, which simulate the splash load in marine environment, are shown to be more notable in composite facings with orientation of lay-ups that are resin dominated (for example, $[\pm 45]_{2S}$). The seawater absorption in the carbon fiber/vinylester composite leads to several deleterious effects resulting in new failure modes and deterioration in long-term durability.

Keywords Carbon fiber · Multiscale · Durability · Seawater · Temperature · Fatigue

D. Penumadu (✉)
CEE Department, University of Tennessee, 325 John D. Tickle Building, 851 Neyland Drive, Knoxville, TN 37996-2313, USA
e-mail: dpenumad@utk.edu

P. Davies and Y.D.S. Rajapakse (eds.), *Durability of Composites in a Marine Environment 2*, Solid Mechanics and Its Applications 245,
https://doi.org/10.1007/978-3-319-65145-3_4

1 Introduction and Materials

Marine composites are subjected to sea environment and thus are exposed to seawater for sustained periods of time. Due to the recent weather changes, naval ships now have access to cold regions for longer durations and thus understanding the coupled effects of temperature and seawater are also important on the static, fatigue, and fracture behavior of the fiber-reinforced polymer composites. This chapter presents a recent summary of the author's work on the effects of sea environment on the mechanical response of carbon fiber composites and sandwich structures. The face sheet materials of interest for US Navy include T700 carbon fiber-based fabric (12k tows and plane weave) and vinylester resin (Derakane 510A) and typically employs VARTM infusion process with a targeted post-cure protocol for optimal properties. Vinylester resins are especially of interest due to their low absorption of moisture and cost. Large panels of approximately 1.5 × 1.5 m were manufactured and coupon samples are obtained from various orientations. To evaluate the effects of long-term exposure to seawater, coupon samples were soaked at 40 °C for an extended period of time until full saturation state was reached through the diffusion process, confirmed by the periodic evaluation of weight gain measurements with time.

The composite sandwich panels were fabricated using the VARTM process (Fig. 1) with the facing material made of carbon stitch bonded fabric designated as LT650-C10-R2VE supplied by the Devold AMT AS, Sweden. This was an equi-biaxial fabric produced using Toray's Torayca T700 12k carbon fiber tow with a vinylester compatible sizing. The weight of the fabric was 634 g/m^2 with 315 g/m^2 of fiber in the 0° direction and 305 g/m^2 in the 90° direction. Both directional fibers were stitched with a 14 g/m^2 polyester knitting thread. Toray's Torayca T700 carbon fiber was chosen because of its lower cost and higher strength. The T700 fiber had a vendor reported tensile strength of 4.9 GPa, a tensile modulus of 230 GPa, and an elongation of 2.1%. The matrix used was Dow Chemical's DERAKANE 510A-40, a brominated vinylester, formulated for the VARTM process. The bromination imparts a fire-resistant property to the composite. The fiber volume fraction was found to be 58–60% including the polyester stitch fiber. The core material used was H100 Divynicell PVC foam having an average density of 100 kg/m^3 and average cell size of 0.15 mm.

2 Microstructure and Multiscale Effects

An example optical micrograph of a composite facing sample that is polished is shown in Fig. 2. The nominal diameter of the carbon fiber reinforcement is 7 μm. As can be seen from this figure and values from Table 1, the arrangement of fibers in the composite shows a typical variation of zero (touching) to four fiber diameters as the spacing between the fibers. Thus in order to understand the degradation

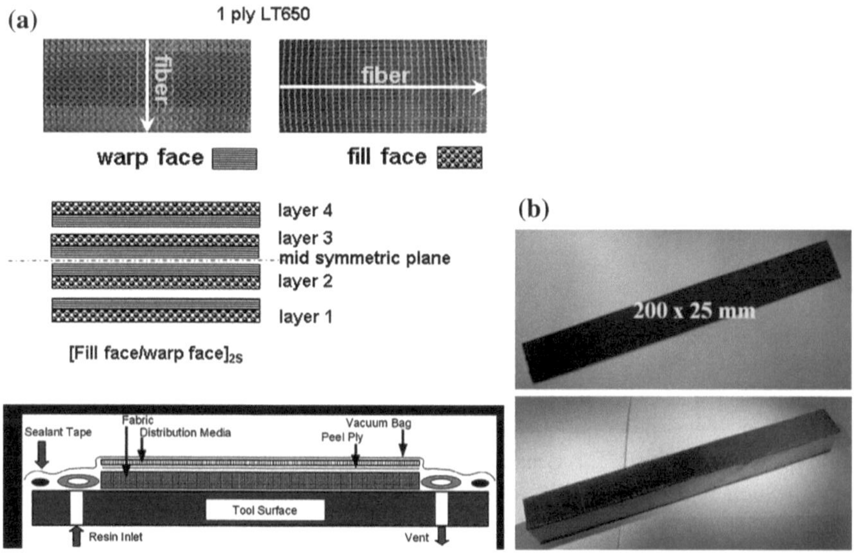

Fig. 1 **a** Biaxial fabric lay-up and VARTM setup, **b** facing and sandwich sample

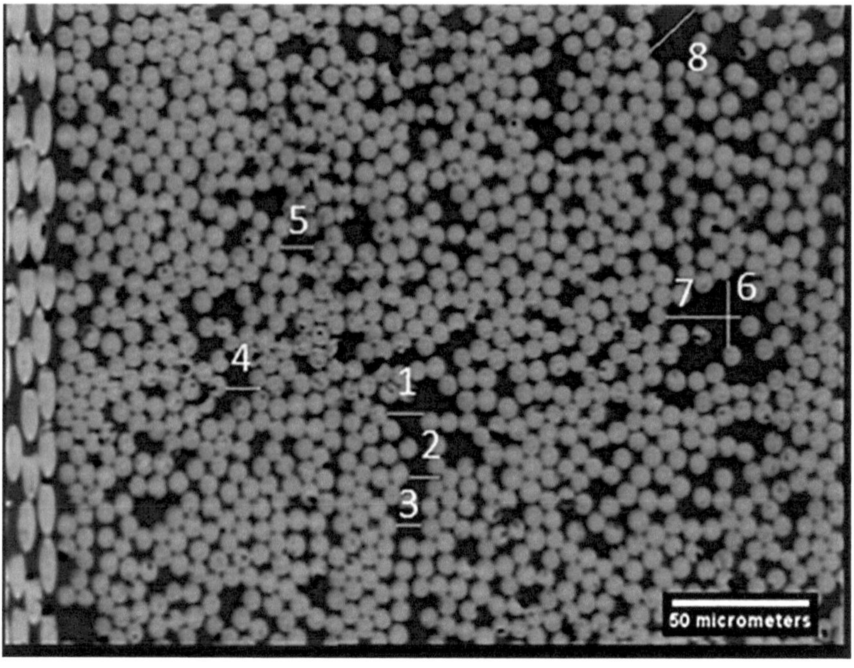

Fig. 2 Microstructure of carbon fiber vinyl ester composite

Table 1 Fiber arrangement within a bundle corresponding to microstructure

Fiber to fiber distance	
Measurement	Length (um)
1	12.919
2	10.526
3	8.612
4	12.44
5	11.483
6	22.967
7	26.794
8	21.339

resulting from the exposure to seawater, one needs to understand the effects at a single fiber scale, potential interface degradation, effects at tow level, and then composites accounting for considered manufacturing process (VARTM for this study). Such multiscale studies comprehensively addressing the durability effects from marine environment are nonexistent to date and this chapter partially addresses the approaches suitable for such comprehensive science based approach to provide a basic understanding of this very complex degradation phenomenon.

A fundamental understanding of degradation associated with harsh sea environment on the mechanical behavior of individual carbon fibers is also essential. Since carbon fiber is typically made of turbostratic graphite crystals, amorphous region, and pore spaces in between, and the fibers are applied with surface chemical treatment called sizing, to optimize the fiber surface to interface with matrix resin and also help with the handling of the fiber during its manufacturing processes, exposure to moisture and seawater can have noticeable effects. Often in the past literature, this aspect is neglected with the assumption that carbon fibers are homogeneous and do not degrade with exposure to moisture, largely a misconception to date. Studying the mechanical properties of single carbon fiber is very challenging requiring very sensitive load and deformation sensing system [1] and the existing standards [2, 3] do not adequately provide a methodology for reliable mounting and testing of individual fibers accounting for system compliance considering the nonlinear elastic response of carbon fibers. In this study, the exposure to sea environment over an extended period of time was simulated in the laboratory by combining the soaking tests of fiber bundles in seawater at elevated temperatures. Seawater absorption and desorption and coupled temperature cyclic effects may degrade the properties and performance of sized carbon fibers. A nano-tensile testing system (Fig. 3) was used to evaluate the mechanical properties of single carbon fibers with diameters around 7 μm.

The unique testing system employed in the study has a load and displacement resolutions comparable with that of a nano-indentation system, i.e., less than 50 nN in axial force and 1 nm resolution in axial displacement using capacitance plates. In addition to exceptional quasi-static performance, simultaneous dynamic loading can

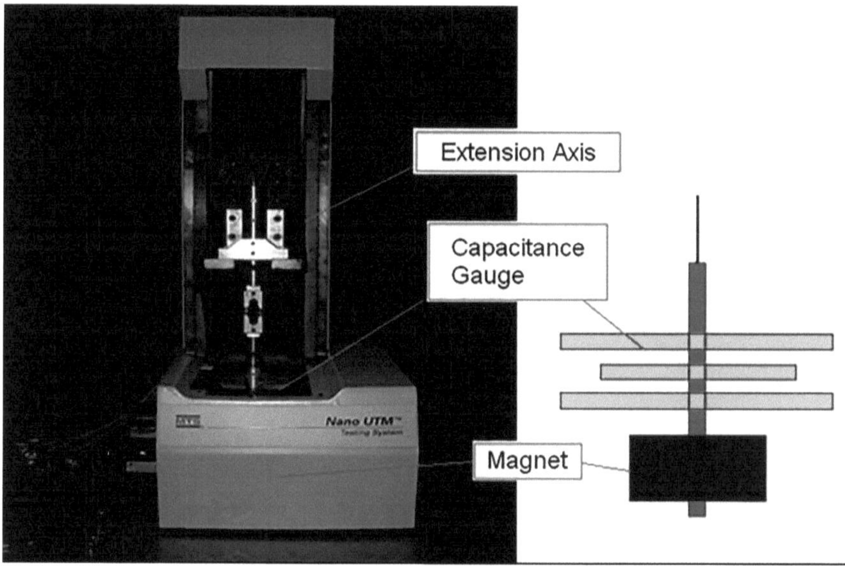

Fig. 3 Single carbon fiber testing system for precise studies on the effect of seawater at single carbon fiber scale

be applied during typical tensile extension of a single fiber, thus obtaining the dynamic elastic moduli (storage and loss modulus values) as a function of the globally applied strain of the fiber sample, instantaneously capturing even small mechanical property changes often unseen in the traditional quasi-static testing approach.

These features provide unique opportunity to measure nonlinear stress/strain response, even in the pre-failure state, an important aspect of most carbon fibers. The stress–strain behavior of individual carbon fibers obtained from Devold LT650 T700 stitched fabric was evaluated using the above universal testing machine equipped with a nano-mechanical actuator. Fibers were soaked at 40, 60 and 80 °C for 8 weeks with extractions done weekly for testing. All samples were then desiccated before testing, using a custom mounting procedure developed by the author's group [1]. In addition, reference (control) samples were placed in a desiccator for time aging studies and with fibers soaking in seawater for target duration. Environmental degradation of the mechanical properties of single carbon fibers was observed as shown in Fig. 4. The fibers that were kept in the desiccator for a target duration corresponding to fibers being aged at target temperatures is termed 'Ambient and Aged' in Fig. 4 The storage modulus, SM, is the instantaneous elastic response of the sample, and is ideal for comparing carbon fiber samples because of the inherent nonlinearity in their stress/strain response. Strong evidence has been presented here based on experimental data to support the notion that carbon fibers are not inert to seawater exposure, and the mechanical degradation was demonstrated using the reduction in storage modulus values as a function of elastic strain

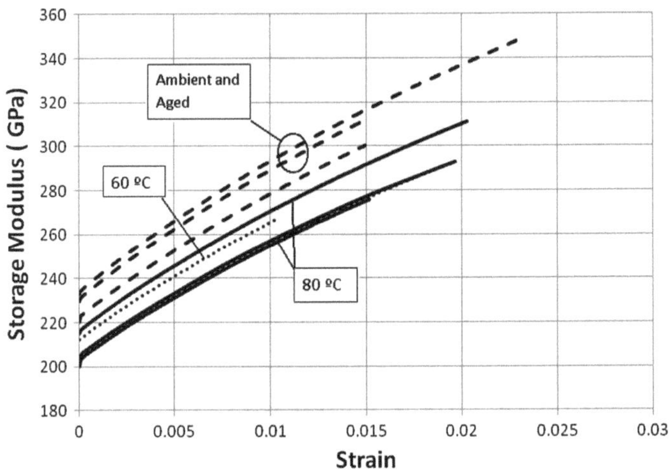

Fig. 4 Effect of seawater at elevated temperature on elastic behavior of single carbon fibers extracted from the devold LT650 T700 biaxial fabric

levels. Most of the literature assumes that carbon fibers individually are not affected by fluids, and this is not a valid assumption.

To accelerate the aging process, fiber properties were obtained from seawater baths maintained at 40, 60, and 80 °C. Work presented here simply demonstrates elastic stiffness loss with time, temperature, and seawater exposure. It would be natural to assume that other related mechanical property degradations would also occur at a single fiber level, including failure stress state and fatigue analysis, a topic of future study for the author's group. It is equally important to learn the degradation mechanism so as to create or improve fiber properties and improved interfaces with target resin to be more effective in resulting composite materials with less degradation due to harsh sea environment. Since the tensile failure of single fibers is generally considered to follow Weibull statistics, a large number of fibers need to be considered in the experimental campaign for reliable interpretation on the failure strength dependency as a function of seawater exposure for accelerated conditions. However, the pre-failure response generally does not vary significantly for various filaments within a tow of carbon fibers.

3 Monotonic Tensile Behavior and Seawater Effects

Composite facing specimens of carbon fiber vinylester (CF/VE) were prepared using different ply lay-ups, namely $[0/90]_{2S}$ and $[\pm 45]_{2S}$ to achieve fiber and matrix dominated failure respectively. All tensile tests were performed under deformation control at a constant crosshead rate of 0.1 mm/min, at room temperature in air. In order to minimize the effect of gripping stress on the mechanical behavior and to avoid failures within tab section, 25 mm long tabs were attached to the ends of the specimens using a suitable adhesive with precise thickness control for the adhesive to ensure alignment of top and bottom parts of the specimens inside hydraulic grips. When experimentally feasible, an extensometer was used to record strain data for comparison purposes. It showed comparable results with Digital Image Correlation (DIC) technique used to obtain surface strain data. The DIC system (commercially available, VIC-3D) uses dual-cameras to measure the shape of an object, surface displacements, and full-field surface strains in three dimensions. The basic method tracks the gray value pattern in small pixel neighborhoods, called subsets, during deformation. The experimental setup is shown in Fig. 5 which includes a biaxial mechanical testing system and 3D-DIC equipment and example rendering of 30 pixel × 30 pixel tracking regions for surface displacement based interpretation of spatially resolved strains.

Figures 6 and 7 depict damage localization and evolution at identified stress levels for matrix dominated specimens. Surface 3D DIC principal axial strains clearly show the location of strain localization and damage accumulation of CF/VE $[\pm 45]_{2S}$ specimens and with large variation in spatially resolved strain values for tensile stress equal to 100 MPa. For axial stress values in the range of 20–80 MPa, corresponding to elastic range, damage evolution in the form of axial strain experienced was distributed along the specimen gage length. Increasing to a stress level at 100 MPa results in a sudden growth of localized strain (shear) bands. However, a completely different damage scenario was observed for fiber dominated CF/VE $[0/90]_{2S}$ specimen displaying localization zones from early stages of applied external stress starting at 20% of failure stress. With stress increase from 200 to 1000 MPa, shows the additional distribution of localized damage and strain

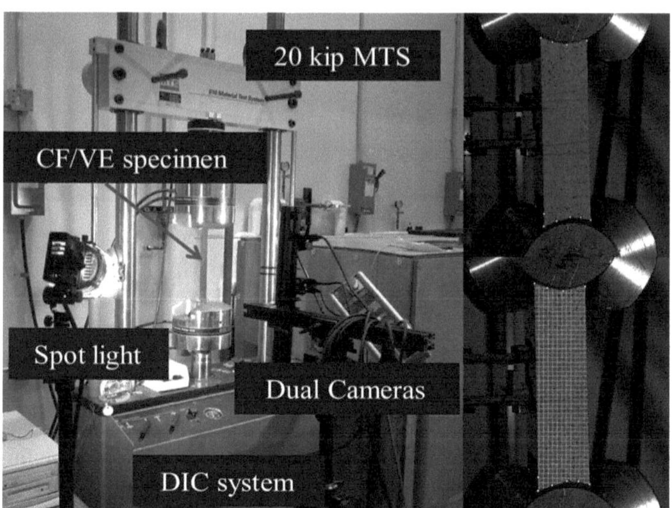

Fig. 5 Servo-hydraulic testing system and 3-D digital image correlation setup

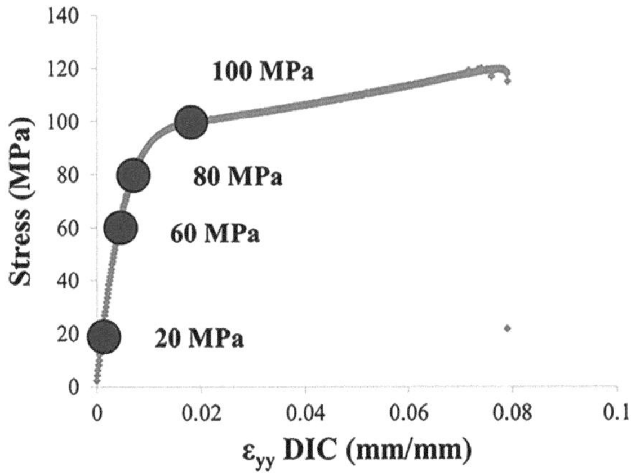

Fig. 6 Tensile stress–strain behavior of matrix dominated specimen ($[\pm45]_{2S}$)

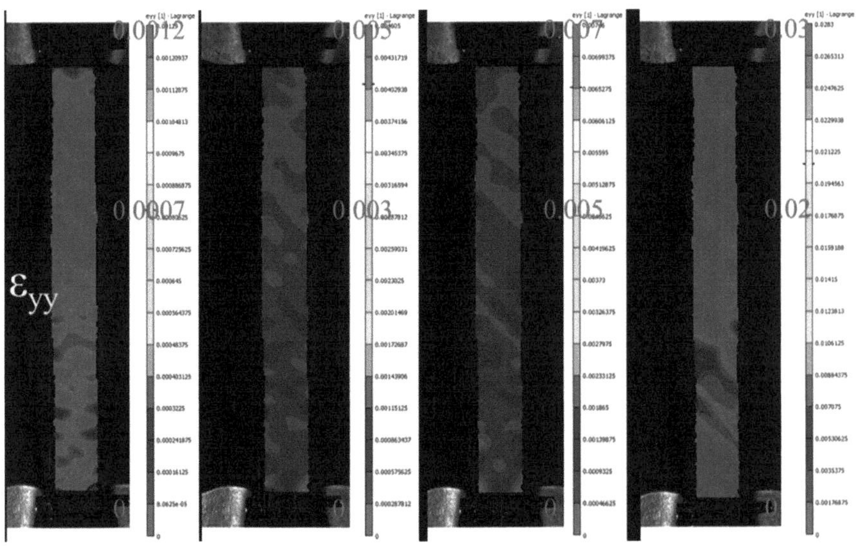

Fig. 7 Axial strain localizations for four tensile stress values indicated in Fig. 6 for the matrix dominated composite specimen ($[\pm45]_{2S}$); 20 MPa on the *left* to 100 MPa on the *right*

accumulation eventually leading to catastrophic failure. The goal of these experiments was to evaluate relative effects of seawater exposure on monotonic behavior considering localized response.

CF/VE samples were also subjected to controlled rate of strain loading with ten tensile load–unload loops (corresponding to unloading at 20, 40, 60, 80 MPa, and 5 loops at 100 MPa) to evaluate accumulated plastic strain as shown in Fig. 8. Fiber dominated CF/VE $[0/90]_{2S}$ composites under ten cycles of up to 80% failure stress

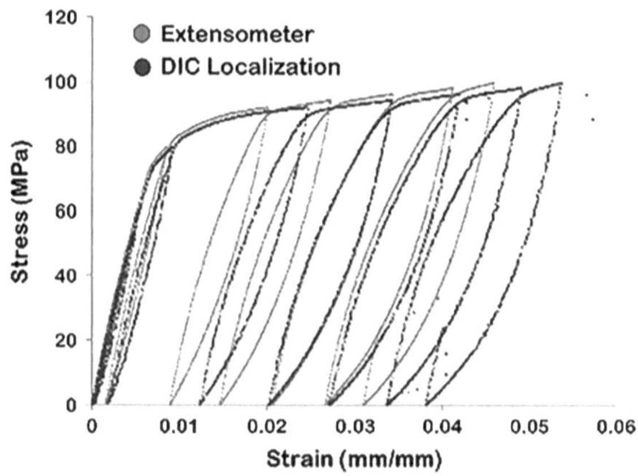

Fig. 8 Monotonic tensile stress strain relation with unload–reload loops for $[\pm 45]_{2S}$

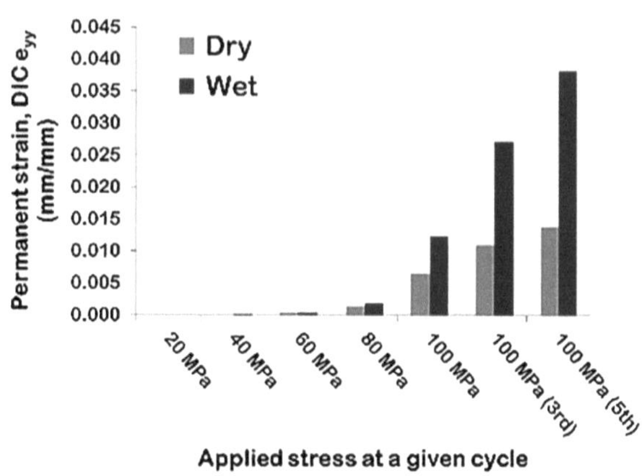

Fig. 9 Accumulated plastic strains for $[\pm 45]_{2S}$ specimen and the effects of long-term seawater exposure contributing to augmented damage

load–unload did not show considerable accumulated permanent strain; therefore, this part of the study focused only on matrix dominated CF/VE. In order to evaluate the damage evolution due to seawater effects compared to time-aged sample, a saturated CF/VE $[\pm 45]_{2S}$ that was soaking in seawater for 3 years was also subjected to similar testing. This particular sample had a moisture uptake of 0.8% based on periodic weight gain measurement. Results included in Figs. 8 and 9 show that exposure to seawater corresponding to *wet sample* substantially increased the amount of plastic strain (referred as damage in this chapter) when compared to *dry sample*, evaluated at various strain or stress levels, quantifying the relative effect of marine environment on damage evolution. After 10 cycles of tensile load–unload cycles, saturated $[\pm 45]_{2S}$ CF/VE sample showed substantial matrix cracking on the sample surface unlike the case of dry and aged specimen of $[\pm 45]_{2S}$ CF/VE composite. Corresponding to the 10th load–unload cycle at the amplitude of axial stress prior to unloading of 100 MPa, large strain variations along the sample length with high strain localization at mid height were also observed. The advantage of

DIC monitoring is that spatially resolved strain data can be carefully observed for evaluating local structure and relation to observed surface strain variations; on the other hand, using only extensometer one could overlook the damage evolution as seen above. Approximately 20% difference was observed for DIC-based axial strain (ε_{yy}) at the end of 10th cycle compared to extensometer based measurements. Damage accumulation zones identified for such tests for various specimen locations are being further studied using high-resolution X-ray tomography to elucidate microstructure role further.

4 Fatigue Behavior of Facing Due to Sea Environment

Under tension–tension fatigue test, orientation dependence of CF/VE exhibit disparate failure mechanisms only when fatigued in the air or under immersed conditions that lead to earlier failures when submerged in fluid. It was shown [4–7] in the past that secant and tangent modulus, E_S and E_T, respectively, were used to determine the degradation due to seawater confinement effect at each orientation. The fatigue life of composites is denoted by the number of load cycles to fatigue failure. It was observed when comparing peak strain subjected to the same stress level of $0.5\sigma_{failure}$ between fiber and resin dominated composite laminates, seawater confinement did not impact fatigue life on fiber-dominated specimen. On the other hand, seawater confinement effects significantly reduced fatigue life on resin dominated CF/VE. Tangent moduli evolution as a function of number of cycles of loading as a basis for comparison between fiber and matrix dominated CF/VE at the same stress level of (0.67 $\sigma_{failure}$) showed similar observations. Therefore, only matrix-dominated samples due to sea environmental effects are focused in this study. Cyclic fatigue tests were suspended when the applied number of cycles reached one millions cycles. This fatigue endurance limit was used in several studies in the past [6, 8, 9].

CF/VE composite samples with deformation response that is matrix dominated ($[\pm 45]_{2S}$), had an average static tensile strength of 120 MPa. Fatigue life for these matrix dominated facings were obtained at varying maximum applied stress levels of 0.8, 0.67, 0.6, and 0.5. Fatigue performance was obtained for these samples with either air or seawater as the confining medium while the specimen was subjected to cyclic tensile loading at an R-ratio of 0.2 at a frequency of 1 Hz as shown in Fig. 10. This figure shows the configuration for simulating one-sided exposure to seawater while the sample is subjected to fatigue loading. A thin latex member provides the access to the deforming specimen to seawater under cyclic loading as shown. A modified setup for extensometer which uses 100 mm gage length is employed as shown in this figure. For a fatigue sample with no confining fluid, it corresponds to fatigue loading in the air, which serves as a reference for changes due to the presence of seawater corresponding to degradation in fatigue life. One sided confinement was considered as the sandwich structures employed in ship structures are typically only exposed to seawater on the outer surface.

Fig. 10 One-sided seawater confinement/exposure for $[\pm 45]_{2S}$ specimen and tension–tension (R = 0.2, 1 Hz) cyclic fatigue setup, 100 mm gage length corresponds to distance between knife edges of extensometer on the gage section

Complete confinement all around the sample corresponds to worst case scenario for maximum degradation condition, though unrealistic in practice.

Fatigue life of matrix-dominated $[\pm 45]_{2S}$ CF/VE samples due to complete seawater confinement and one-sided confinement/exposure effects are summarized in Fig. 11. It can be clearly seen from this figure that scatter in the results for fatigue life for repeating samples ($\sim$6 in this study) is relatively large (as expected) for a target maximum cyclic stress amplitude [10]. The mean response indicates clearly that the fatigue life of these composites is much higher when they were confined in the air as compared to seawater confinement. For maximum stress amplitude of 50% of the static tensile strength, fatigue life exceeded a million cycles (fatigue limit), thus experiments below that stress level were not considered. As the maximum cyclic stress amplitude increased from 0.5 to 0.8, relative effects of water confinement also decreased. Figure 12 shows the relative changes in fatigue life for maximum cyclic stress amplitude corresponding to a value of two-thirds of its static tensile strength. Wet sample corresponds to fully saturated condition based on weight gain observations. Dry sample corresponds to time aged composite in low humidity environment while wet samples were soaked to reach saturation weight gain condition, typically 10 weeks or longer. As can be seen, a reduction in fatigue life of nearly 50% is expected for ship structures in service for a long time allowing outer facing of sandwich structure to be saturated and subjected to fatigue with water available on one side.

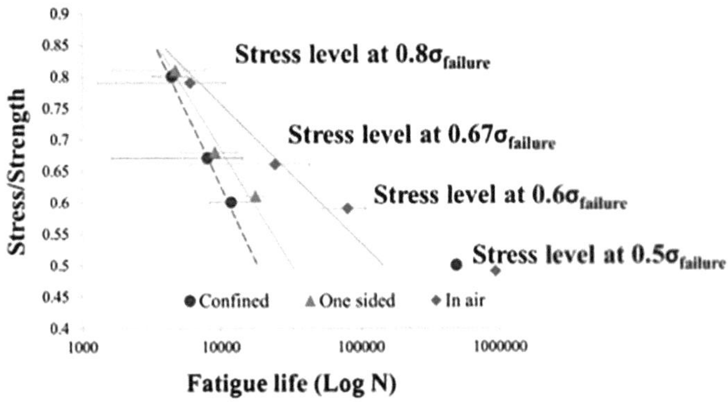

Fig. 11 Seawater effect on master curves for fatigue life

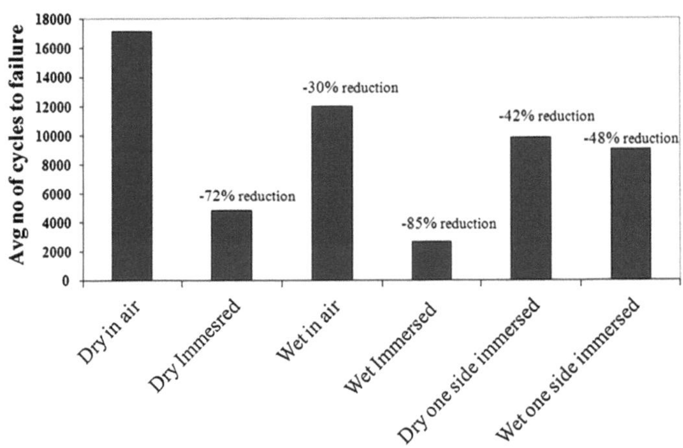

Fig. 12 Seawater effect on tension fatigue life at a stress level of ($0.67\sigma_{\text{failure}}$)

The large degradation resulting from the availability of seawater while the fatigue specimen is undergoing repeated loading was due to a new failure mode which was confirmed using high resolution computed tomographic analysis of specimens after failure. Specimens fatigued in air showed damage throughout the cross-section of the failed zone while those confined with seawater largely failed from delamination of outer lay-up. The matrix cracking during cyclic fatigue loading allows incompressible water into these cracks and result in very large stresses on the outer part of the facing during downward loading, resulting in premature failure and not allowing the entire cross-section of the material to contribute to fatigue strength. Prediction of fatigue life in composites is more complicated as compared to metals because the failure mechanism involves several factors including fiber breakage, matrix cracking de-bonding, and the complex role of interface/interphase. With different damage mechanisms and related accumulation in composites, experimental results are a must for a given material system and manufacturing conditions. Observing the secant modulus degradation of facing material during the service of the sandwich ship structure can serve as a good indicator for the expected fatigue life remaining for service considerations.

5 Role of Interface in Sandwich Structures

Interfaces at multiple scale play a very important and complex role in capturing the degradation in mechanical behavior of reinforced composites subjected to harsh environment associated with naval structures including seawater, temperature, and hydrostatic pressure. At a fiber level, chemical sizing can absorb moisture depending on the chemistry and storage conditions prior to use of the material form (unidirectional pre-preg, stitched fabric) and can significantly reduce the interfacial shear strength. The relative loss can be quantified using single fiber fragmentation testing, but the challenge is to mimic the state of the cured resin system in single fiber composite to be representative of the manufactured large scale composite. However, relative effects can be obtained precisely using laboratory scale studies and can also be effectively utilized to target fiber sizing that is more robust for low cost and high volume applications such as naval composites. The interface between facing and foam core also plays a very important role and its degradation is not considered in the literature in depth. Table 2 summarizes critical energy release rates obtained from DCB specimens (Fig. 13).

A picture of a small sample of sandwich structure material considered in this study and the corresponding X-ray computed tomography [11] image of its cross-section at one location is shown in Fig. 14. The complex interface between the face sheet and foam core material can be clearly seen from the reconstructed X-ray tomographic projection. Despite the fact that the surface of the H100 Divinycell foam core has access to its cells for the vinylester resin to infuse and develop an intimate bond, a number of void regions at the interface are visible, and the jagged edge of the resin front is quite rough with sharp transitions. Our detailed work has shown that when this resin rich interface is exposed to seawater, the interface degrades substantially, more so than the facing material itself, and results in delamination crack propagating within the matrix region. For dry interface, delamination experiments indicate interface cracks always propagating into the lower density foam region. Thus, a substantial degradation of approximately a third

Table 2 Delamination testing of face sheet and foam core summary results

Condition (no. of samples)	Critical energy release rate, G_c (N/m^2)	Degradation
Dry (15)	541–963	
Wet (10)	451–632	30%

Fig. 13 DCB specimen testing setup for interface delamination studies that shows cryogenic load cell

Fig. 14 Picture of a CF/VE-H100 PVC sandwich material and X-ray-computed tomography projection showing the interface between composite face sheet and cellular core

in the reduction of critical energy release rate was observed from the double cantilever beam specimens as noted in Table 2. This observation leads to an important aspect of optimizing surface of core materials with the resin system being used to potentially lessen the impact of sea environment on interfacial delamination toughness.

6 Summary and Conclusions

A fundamental study to evaluate the environmental degradation associated with long-term exposure on mechanical properties of carbon fiber-reinforced vinylester facings at different orientations and various specimen sizes was implemented. The tension–tension fatigue effects are shown to be more notable in composite facings with orientation of lay-ups that are resin dominated (for example, $[\pm 45]_{2S}$). The moisture absorption in the carbon fiber/vinylester composite leads to several deleterious effects resulting in new failure modes and deterioration in long-term durability. Specifically, it has been demonstrated that moisture/seawater has impact on the following:

(1) A gradual reduction in nonlinear elastic properties of single carbon fibers was observed for T700 fibers when they were subjected to aggressive soaking conditions. Since the behavior was measured based on very precise nano-tensile testing, this demonstrates that the amorphous and porous portions of the single carbon fiber microstructure can cause degradation and should be considered in the future studies. Carbon fibers should not be considered as inert, as is often assumed, for marine environment.

(2) Under monotonic tensile loading conditions, the damage evolution for seawater saturated composite facings that are matrix dominated was much more than dry specimens. Damage was quantified based on accumulated plastic tensile strain and was spatially resolved using three-dimensional digital image correlation technique. Both dry and saturated specimens showed stain localizations well below the eventual failure stress and likely results from variations in local fiber volume and resin fraction and fiber orientation.

(3) Tension–Tension fatigue life of carbon fiber vinylester composites is much higher when they were fatigued in the air as compared to conditions associated with seawater confinement. For maximum stress amplitude of 50% of the static tensile strength, fatigue life exceeded million cycles (fatigue limit).

(4) Specimens fatigued in air showed damage throughout the cross-section of the failed zone while those confined with seawater largely failed from delaminations of outer lay-up. The matrix cracking during cyclic fatigue loading allows incompressible water into these cracks and results in very large stresses during downward loading leading to premature failure.

(5) Interfaces at multiple scales play a very important and complex role in capturing the degradation in mechanical behavior of reinforced composites subjected to harsh marine environment, leading to reduced fiber-matrix interfacial shear strength and lower delamination toughness between face sheets and foam core materials.

Acknowledgements The author would like to acknowledge the support of the Solid Mechanics Program of the US Office of Naval Research through ONR Award No: N00014-16-1-2765 and under a program managed by Dr. Yapa Rajapakse and is gratefully acknowledged. The contributions of former graduate students of the author, Drs. Akawat Siriruk, and Matthew Kant, are very much appreciated.

References

1. Kant M, Penumadu D (2014) Dynamic mechanical characterization for nonlinear behavior of single carbon fibers. Compos Part A-Appl Sci Manuf 66:201–208
2. ASTM C1557-03 (2008) Standard test method for tensile strength and Young's modulus of fibers
3. ISO 11566:1996 (1996) Carbon fibre—determination of the tensile properties of single-filament specimens
4. Guedes RM (2007) Durability of polymer matrix composites: Viscoelastic effect on static and fatigue loading. Compos Sci Technol 67(11–12):2574–2583
5. Hwang W, Han KS (1986) Fatigue of composites—fatigue modulus concept and life prediction. J Compos Mater 20(2):154–165
6. Lee LJ, Yang JN, Sheu DY (1993) Prediction of fatigue life for matrix-dominated composite laminates. Compos Sci Technol 46(1):21–28
7. Petermann J, Schulte K (2002) Effects of creep and fatigue stress ratio on the long-term behaviour of angle-ply CFRP. Compos Struct 57(1):205–210
8. Yang JN, Lee LJ, Sheu DY (1992) Modulus reduction and fatigue damage of matrix dominated composite laminates. Compos Struct 21(2):91–100
9. Lee LJ, Fu KE, Yang JN (1996) Prediction of fatigue damage and life for composite laminates under service loading spectra. Compos Sci Technol 56(6):635–648
10. Siriruk A, Penumadu D (2014) Degradation in fatigue behavior of carbon fiber-vinyl ester based composites due to sea environment. Compos Part B-Eng 61:94–98
11. Siriruk, A, Penumadu, D and Thomas K (2015) Damage evolution in carbon fiber vinyl ester marine composites and sea water effects. In: 20th international conference on composite materials, ICCM-20, Copenhagen

Statistical Long-Term Creep Failure Time of Unidirectional CFRP

Yasushi Miyano and Masayuki Nakada

Abstract A method for statistical prediction of the long-term creep failure time of CFRP using the statistical static strengths of CFRP at various temperatures and the viscoelasticity of matrix resin is proposed based on Christensen's model of viscoelastic crack kinetics. The tensile strength along the longitudinal direction of unidirectional CFRP constitutes important data for the reliable design of CFRP structures. The authors developed a reliable method for testing creep and fatigue strengths as well as static strength at elevated temperatures for resin-impregnated carbon fiber strands (CFRP strands) as unidirectional CFRP. Two kinds of CFRP strands with two types of PAN-based carbon fibers with high strength and high modulus were examined on the viewpoint of failure mechanism. The statistical static strengths of these CFRP strands and the creep compliances of matrix resins were measured at various temperatures. The tensile creep failure times of these CFRP strands are predicted statistically based on a prediction method using measured data. The predicted creep failure times of these CFRP strands were compared with the creep failure times of these CFRP strands measured experimentally and statistically. Additionally, the statistical temperature-dependent static strengths are also discussed for CFRP strands of two types of pitch-based carbon fibers with low modulus and high modulus.

Keywords CFRP · Viscoelasticity · Life prediction · Tensile strength · Creep failure time

1 Introduction

The mechanical behavior of matrix resin of CFRP exhibits time and temperature dependence, called viscoelastic behavior, not only above the glass transition temperature T_g, but also below T_g. Consequently, it can be inferred that the mechanical

Y. Miyano (✉) · M. Nakada
Materials System Research Laboratory, Kanazawa Institute of Technology, Hakusan, Japan
e-mail: miyano@neptune.kanazawa-it.ac.jp

P. Davies and Y.D.S. Rajapakse (eds.), *Durability of Composites in a Marine Environment 2*, Solid Mechanics and Its Applications 245,
https://doi.org/10.1007/978-3-319-65145-3_5

behavior of CFRP depends strongly on time and temperature [1–5]. Our earlier reports have proposed the formulation of statistical static, creep, and fatigue strengths of CFRP based on matrix resin viscoelasticity [6, 7].

The tensile strength along the longitudinal direction of unidirectional CFRP constitutes important data for the reliable design of CFRP structures. The authors developed a reliable test method to assess creep and fatigue strengths, as well as the static strength, at elevated temperatures for resin-impregnated carbon fiber strands (CFRP strands) combined with PAN-based high-strength carbon fibers T300-3000 and epoxy resin [8]. Furthermore, the authors confirmed that the time-dependent and temperature-dependent tensile static strength is controlled by the viscoelastic behavior of matrix resin based on the Rosen's shear lag model [9]. Additionally, the authors developed a test method for the CFRP strand of more high-strength carbon fibers T800-12000 and epoxy resin with a highly reliable co-cured tab. The temperature-dependent tensile strength of this CFRP strand was evaluated [10].

Our most recent study undertook the prediction of statistical creep failure time under tension loading along the longitudinal direction of unidirectional CFRP performed using CFRP strands of two kinds of high-strength carbon fibers T800-12000 and T300-3000 and epoxy resin [11, 12]. The statistical creep failure time of these CFRP strands at a constant load and temperature was predicted using statistical results of static tensile strengths of CFRP strands measured at various temperatures, and the viscoelastic behavior of matrix resin. The predicted results statistically agreed well with the experimentally obtained results measured using creep tests for these CFRP strands.

For this study, the proposed method of predicting the statistical creep failure time under the tension loading along the longitudinal direction of unidirectional CFRP from the statistical static strengths of unidirectional CFRP measured at various temperatures is applied to various unidirectional CFRP with carbon fibers of different kinds. First, the method of predicting the statistical creep failure time of CFRP from the statistical static strengths of CFRP measured at various temperatures is proposed again based on Christensen's model of viscoelastic crack kinetics [13]. Second, two kinds of CFRP strands with PAN-based high-strength and high-modulus carbon fibers were prepared as specimens for unidirectional CFRP. Third, the static strengths of these CFRP strands were measured experimentally and statistically at various temperatures. Then the creep failure times of these CFRP strands were predicted statistically using the statistical static strengths based on the predictive method. The creep failure times of these CFRP strands at several constant loads and a temperature were measured experimentally and probabilistically using these CFRP strands for comparison with the predicted ones. Additionally, the predictive method was applied to CFRP strands of two kinds with low-modulus and high-modulus pitch-based carbon fibers. Finally, the time and temperature dependences on the statistical static and creep strengths of various CFRP strands are discussed in the viewpoints of fracture mechanism.

2 Statistical Prediction of Creep Failure Time of CFRP

We have proposed the formulation for the statistical static strength σ_s of CFRP based on the viscoelasticity of matrix resin, as shown in the following equation in our previous paper [7]:

$$\log \sigma_s(P_f, t, T) = \log \sigma_0(t_0, T_0) + \frac{1}{\alpha_s}\log[-\ln(1 - P_f)] - n_R \log\left[\frac{D^*(t, T)}{D_c(t_0, T_0)}\right], \quad (1)$$

where P_f signifies the failure probability, t denotes the failure time, t_0 represents the reference time, T is the temperature, T_0 stands for the reference temperature, σ_0, and α_s respectively denote the scale parameter and the shape parameter on the Weibull distribution of static strength, n_R is the viscoelastic parameter, and D_c and D^* respectively represent the creep and viscoelastic compliances of matrix resin. The viscoelastic compliance D^* for the static load with a constant strain rate is shown as

$$D * (t, T) = D_c(t/2, T). \quad (2)$$

The statistical static strength σ_s is shown by the following equation by substituting Eq. (2) into Eq. (1).

$$\log \sigma_s(P_f, t, T) = \log \sigma_0(t_0, T_0) + \frac{1}{\alpha_s}\log[-\ln(1 - P_f)] - n_R \log\left[\frac{D_c(t/2, T)}{D_c(t_0, T_0)}\right] \quad (3)$$

The relation between the creep failure time and the static failure time can be shown in Fig. 1 [11]. This figure shows the creep strength and the static strength versus the failure time. The creep strength curve is obtainable by horizontally shifting the static strength curve by the amount log A. Therefore, the statistical creep strength σ_c is shown by the following equation.

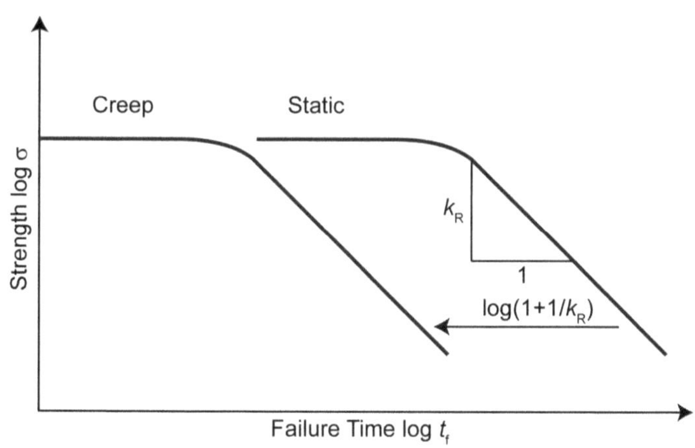

Fig. 1 Time shifting between static strength and creep strength [11, 12]

$$\log \sigma_c(P_f, t, T) = \log \sigma_0(t_0, T_0) + \frac{1}{\alpha_s}\log[-\ln(1-P_f)] - n_R \log\left[\frac{D_c(At/2, T)}{D_c(t_0, T_0)}\right] \tag{4}$$

The failure probability of CFRP under a constant creep stress σ_{c0} can be shown by the following equation from Eq. (4).

$$P_f = 1 - \exp(-F), \log F = \alpha_s \log\left[\frac{\sigma_{c0}}{\sigma_0}\right] + \alpha_s n_R \log\left[\frac{D_c(At/2, T_0)}{D_c(t_0, T_0)}\right] \tag{5}$$

The shifting amount $\log A$ is determined by the slope k_R of the static strength curve shown in Fig. 1 as

$$\log A = \log(1 + 1/k_R), \quad k_R = n_R m_R \tag{6}$$

where n_R is the viscoelastic parameter in Eqs. (1) and (3), and where m_R is the slope of logarithmic creep compliance of matrix resin against the logarithmic time [11, 12].

3 Molding of CFRP Strands

Two kinds of PAN-based carbon fibers and pitch-based carbon fibers of two kinds were prepared in this study, as shown in Table 1. PAN-based carbon fibers are high-strength-type carbon fiber T300-3000 and high-modulus-type carbon fiber M40J-6000 (Toray Industries Inc.). The pitch-based carbon fibers are low-modulus-type carbon fiber XN05 and high-modulus-type carbon fiber XN50 (Nippon Graphite Fiber Co. Ltd.). Weibull distributions of the static strengths σ_c for carbon fibers of four kinds with shape parameters α_c and scale parameters β_c are shown in Fig. 2.

Four CFRP strands with carbon fibers of four kinds and a general purpose epoxy resin jER828 (Mitsubishi Chemical Corp.) were molded using a filament winding system developed by the authors [8]. Actually, 200 specimens for four kinds of CFRP strands are molded, respectively, at one time using this system. Table 2

Table 1 Carbon fibers used in this study and mechanical properties by catalogs

Name	PAN or pitch	Density (g/cm^3)	Tex (g/km)	Elastic modulus (MPa)	Tensile strength (MPa)
T300-3000	PAN	1.76	198	230	3530
M40J-6000	PAN	1.75	225	377	4400
XN05-3000	Pitch	1.65	410	55	1100
XN50-3000	Pitch	2.14	450	527	3830

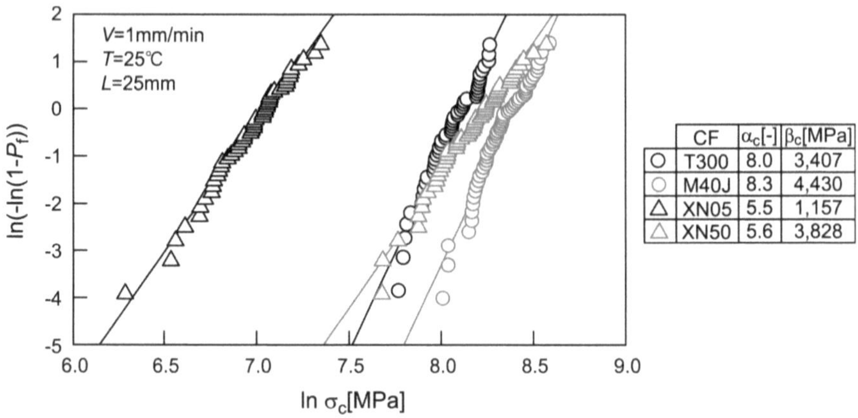

Fig. 2 Weibull distributions of static strengths of single fibers of four kinds

Table 2 Composition and cure schedule of CFRP strand

CFRP strand	Carbon fiber	Composition of resin (weight ratio)	Cure schedule
T300/EP	T300-3000	Epoxy: jER828 (100) Hardener: MHAC-P (103.6) Cure accelerator: 2E4MZ (1)	70 °C × 12 h +150 °C × 4 h +190 °C × 2 h
M40J/EP	M40J-6000		
XN05/EP	XN05-3000		
XN50/EP	XN50-3000		

presents a composition of epoxy resin and the cure condition of CFRP strand, which is the same to those for the Refs. [11, 12]. The gage lengths of CFRP strands are approximately 200 mm. The glass transition temperatures T_g = 160 °C of the epoxy resin are determined from the peak of loss tangent against temperature at 1 Hz using the DMA test. The fiber volume fraction V_f = 55% of CFRP strand is ascertained from the weight of CFRP strands.

4 Creep Compliance of Matrix Resin and Static Strength of CFRP Strands

The dimensionless creep compliance D_c/D_{c0} measured at various temperatures is shown at the left of Fig. 3. Long-term D_c/D_{c0} at T = 120 °C was obtained by horizontally shifting those at various temperatures, as shown at the right of Fig. 3 [11, 12]. The reference temperature and time were selected as T_0 = 25 °C and t_0 = 1 min in this study. The creep compliance at reference temperature and reference time D_{c0} was 0.33 GPa^{-1}. The dashed curve is the dimensionless viscoelastic compliance D^* of the matrix resin under the constant strain rate at T = 120 °C. The maximum slope in this figure is m_R = 0.28 shown in Eq. (6).

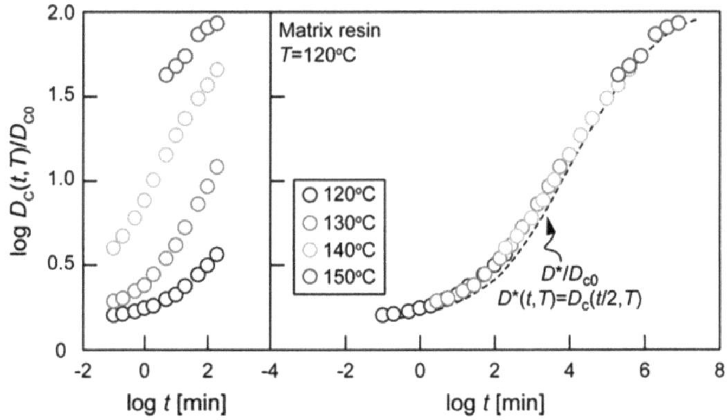

Fig. 3 Dimensionless creep compliance of matrix resin at T = 120 °C [11, 12]

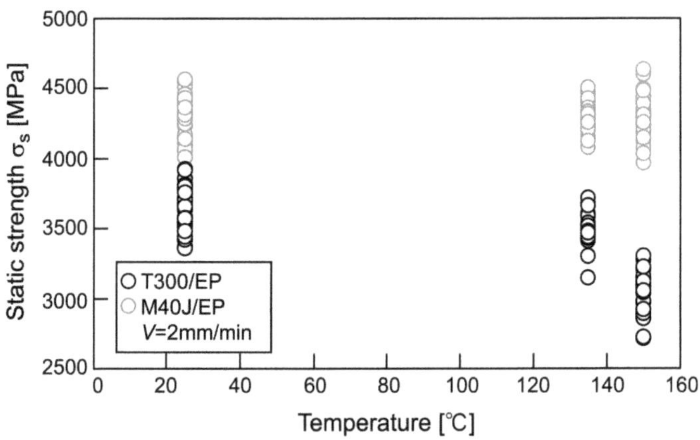

Fig. 4 Static strength of T300/EP and M40J/EP versus temperature

The static tension tests for two kinds of CFRP strands T300/EP and M40J/EP with PAN-based carbon fibers were conducted at three temperature levels, 25 °C, 135 °C, and 150 °C with 2 mm/min cross-head speed. The tensile strength of the CFRP strand σ_s was obtained using the following equation.

$$\sigma_s = \frac{P_{max}}{t_e}\rho \tag{7}$$

Therein, P_{max} stands for the maximum load (N). ρ and t_e respectively represent the density of carbon fiber (kg/m^3) and the Tex of a carbon fiber strand (g/1000 m).

Figure 4 presents static strengths versus temperature for CFRP strands of two kinds: T300/EP and M40J/EP. The static strength of T300/EP decreases with increasing temperature and that of M40J/EP keeps a constant with increasing temperature. The Weibull distributions for the static strength of CFRP strand T300/EP at three temperatures are shown in Fig. 5, where α_s is the shape parameter and β_s is the scale parameter of CFRP strands. Although the scale parameter decreases according to the temperature rise, the shape parameter maintains an almost constant value. Figure 6 is the Weibull distribution for the static strength of M40J/EP at three temperatures. Scale parameter β_s as well as the shape parameter α_s maintain almost a constant value for different temperatures in the case of M40J/EP. Shape parameter $\alpha_s$and the scale parameter β_s at the temperature $T = 25$ °C in these figures can be presumed as shape parameter α_s and scale parameter σ_0 of static strength at the reference temperature $T_0 = 25$ °C and the reference failure time t_0 = 1 min used in Eqs. (1), (3), and (5).

Figures 7 and 8 show the dimensionless static strength σ_s/σ_0 against the dimensionless viscoelastic compliance of matrix resin D^*/D_{c0} simultaneously and temperature for T300/EP and M40J/EP. The relation of σ_s/σ_0 against D^*/D_{c0} can be shown by a straight line with the slope of n_R, which is the viscoelastic parameter in Eqs. (1), (3), and (5). The slopes of both CFRP strands are completely different from one another. The shape parameter α_s of T300/EP is almost equal to that predicted from the shape parameter α_c of carbon fiber T300 shown in Fig. 2 based on Rosen's shear lag model. On the other hand, the α_s of M40J/EP is almost equal to 0 which is completely different from that predicted from the α_c of carbon fiber M40J. It can be presumed that the fracture mechanisms of T300/EP and M40J/EP are quite different. The details of this difference are discussed in the Sect. 8.

All parameters in Eqs. (1), (3), and (5) were determined by measuring the creep compliance of matrix resin and the statistical static strength of CFRP strand at various temperatures through the above process. They are shown in Table 3.

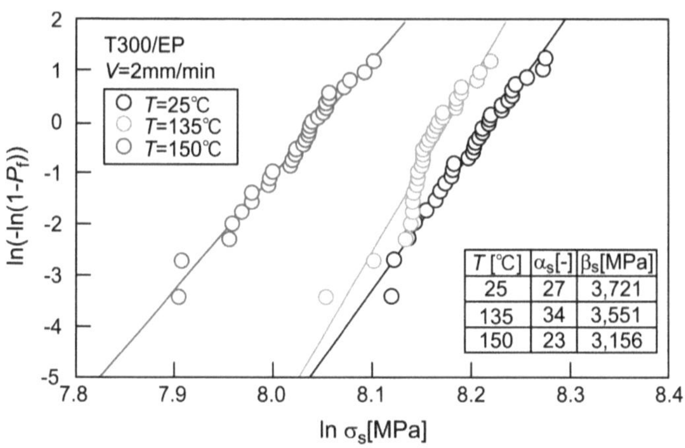

Fig. 5 Weibull distributions of static tensile strength of T300/EP at four temperatures

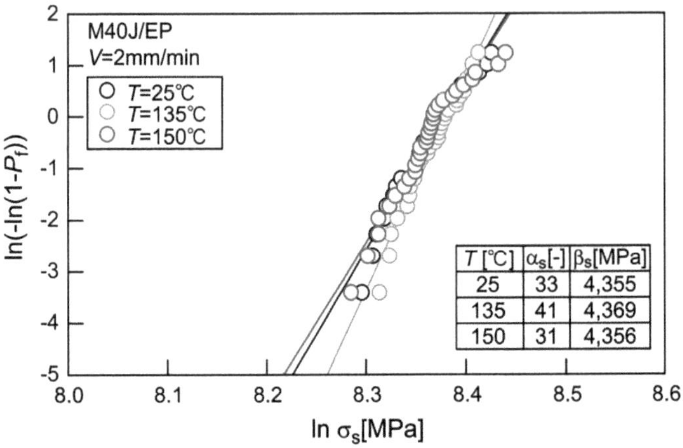

Fig. 6 Weibull distributions of static tensile strength of M40J/EP at three temperatures

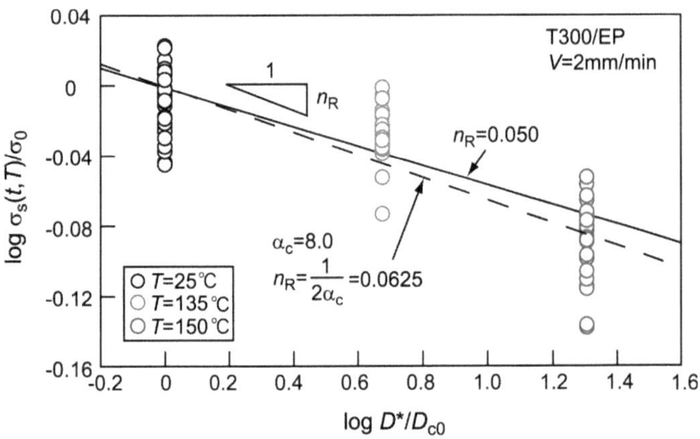

Fig. 7 Static strength of T300/EP against viscoelastic compliance of matrix resin

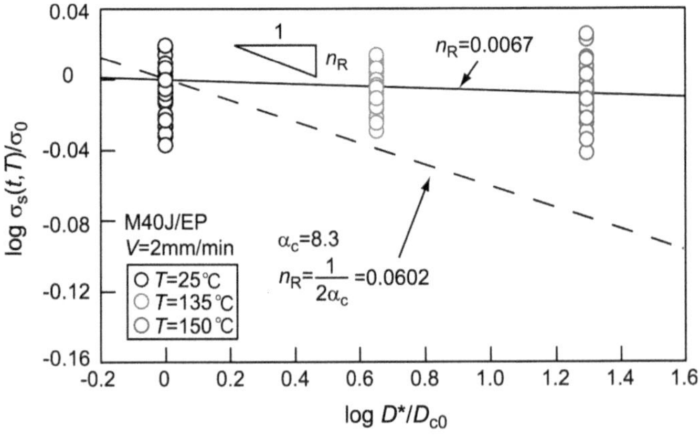

Fig. 8 Static strength of M40J/EP against viscoelastic compliance of matrix resin

Table 3 Parameters for statistical creep failure time prediction

Parameters	T300/EP	M40J/EP
Scale parameter of static strength of CFRP strand at 25 °C: σ_0	3,721 MPa	4,355 MPa
Shape parameter of static strength of CFRP strand: α_s	27	33
Viscoelastic parameter of matrix resin: n_R	0.050	0.0067
Slope of viscoelastic compliance of matrix resin: m_R	0.28	0.28
Slope of static strength of CFRP strand against failure time: k_R	0.014	0.00188
Logarithmic time shifting factor: log A	1.86	2.73

Table 4 Conditions of creep failure tests for CFRP strands

CFRP strand	Temperature T (°C)	Creep stress σ_{c0} (MPa)	σ_{c0}/σ_0 (%)	Number of specimens
T300/EP	120	3,126	84	20
		3,312	89	20
		3,498	94	20
M40J/EP	120	3,658	84	20
		3,876	89	20
		4,094	94	20

5 Creep Failure Time of CFRP Strands

Creep failure tests of T300/EP and M40J/EP were conducted using a specially designed creep failure testing machine [11]. The test conditions are presented in Table 4. Results of the creep failure tests are presented in Figs. 9 and 10. The predicted creep failure probability against failure time calculated by substituting the parameters of Table 3 into Eqs. (5) and (6) is also shown in Figs. 9 and 10. The upper graphs of these figures show the relation of creep strength σ_c against the creep failure time t at $T = 120$ °C, and the lower graphs show the relation of failure probability P_f against the creep failure time t at $T = 120$ °C. The predicted statistical creep failure time agrees well with the experimental data for T300/EP. However, the predicted statistical creep failure time does not agree with the experimental data for M40J/EP.

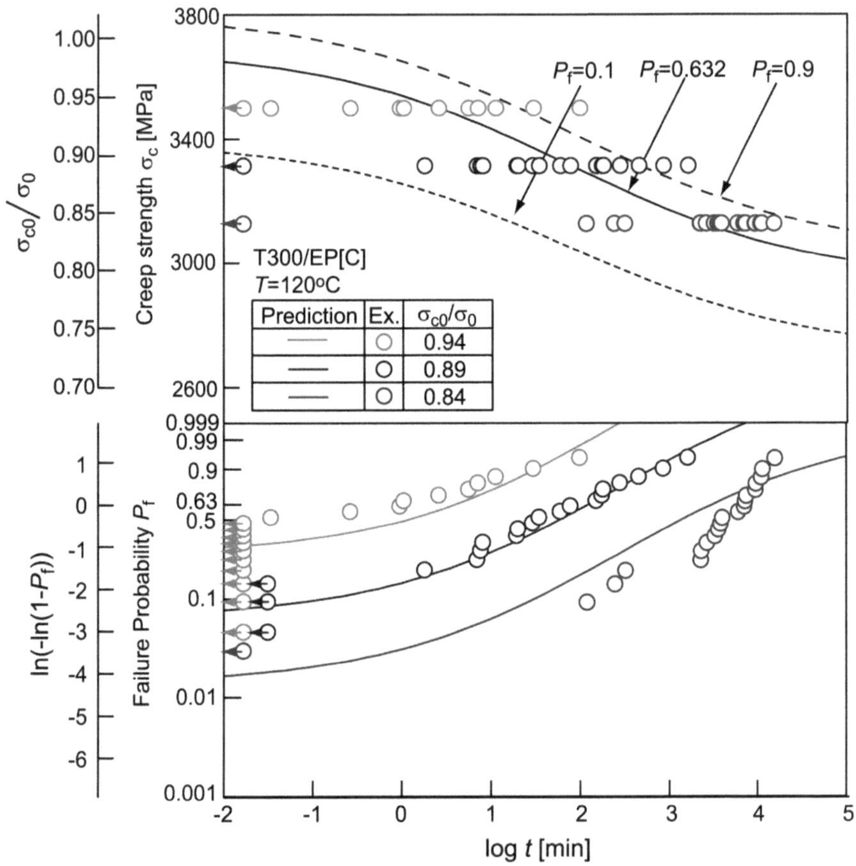

Fig. 9 Failure probability against creep failure time of T300/EP

6 Fractographs of T300/EP and M40J/EP After Static and Creep Tests

Figures 11 and 12 respectively portray fractographs of T300/EP and M40J/EP after the static tests and creep tests. These figures clarify that the fractures of CFRP strands show different tendencies. A markedly uneven failure surface is observed for T300/EP, where penetrations of cracks through fibers are not observed. However, a mirror surface with a wide area is observed in M40J/EP, where the generated cracks penetrate through numerous fibers.

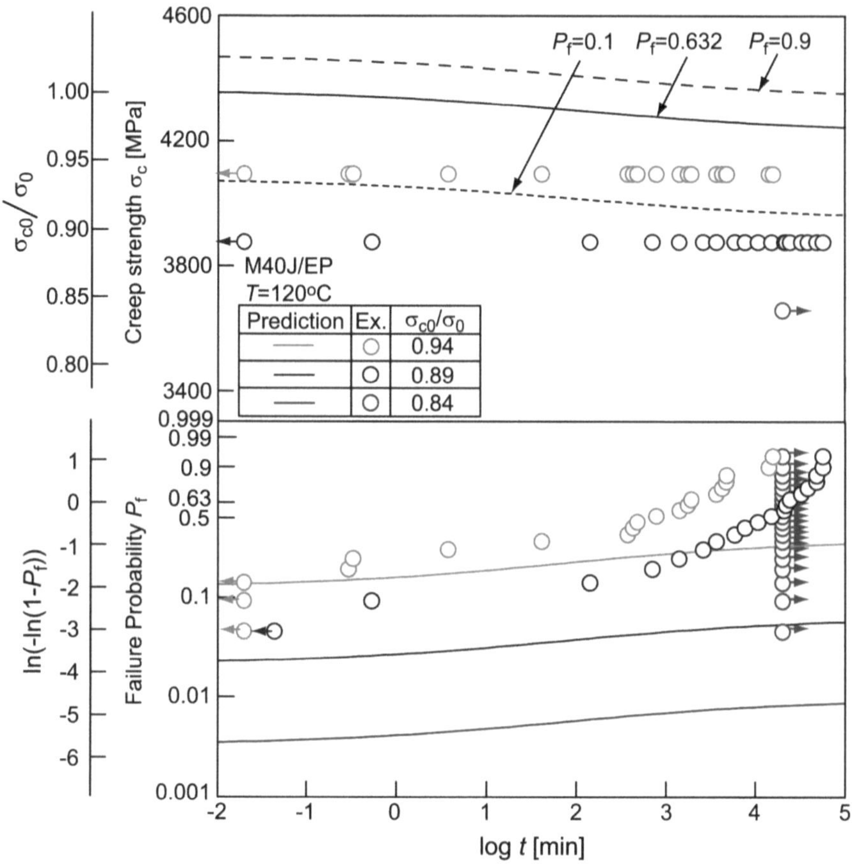

Fig. 10 Failure probability against creep failure time of M40J/EP

7 Temperature Dependence of Static Strength on Pitch-Based Carbon Fiber Strands

Figures 13 and 14 respectively portray the static strengths versus temperature for two kinds of CFRP strands, XN05/EP and XN50/EP, with two kinds of pitch-based carbon fibers (XN05 and XN50; Nippon Graphite Fiber Co. Ltd.) and a general purpose epoxy resin (jER828; Mitsubishi Chemical Corp.). The mechanical properties of XN05 are low modulus and low strength. These properties of XN50 are high modulus and high strength, as shown in Table 1. The static strength of XN05/EP changes clearly with temperature. That of XN50/EP does not change with temperature, as shown in these figures.

Figures 15 and 16 respectively show the static strength σ_s against the viscoelastic compliance of matrix resin D^*/D_{c0} simultaneously and temperature for XN05/EP and XN50/EP. The relation of σ_s against D^*/D_{c0} can be shown as a

Fig. 11 Fractographs of T300/EP and M40J/EP after static tension tests

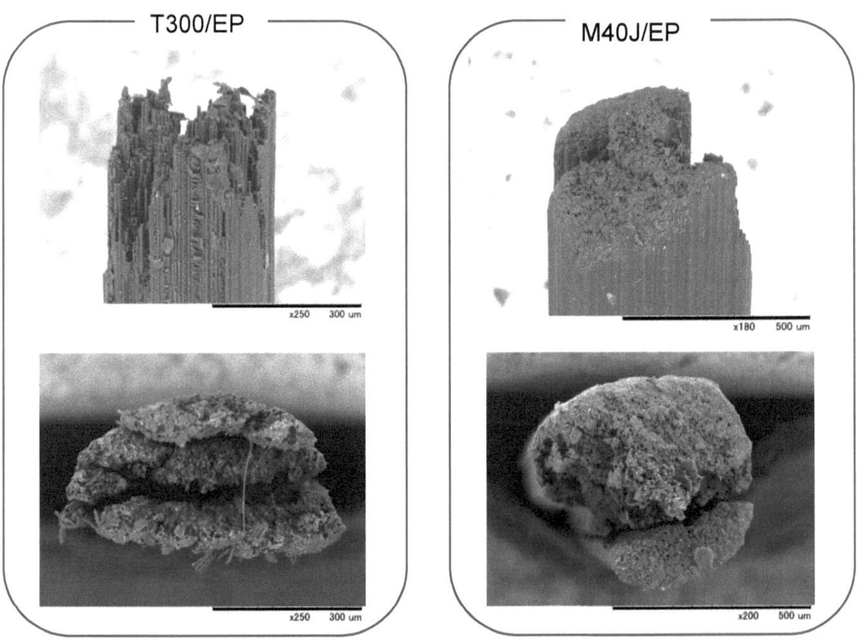

Fig. 12 Fractographs of T300/EP and M40J/EP after creep tests

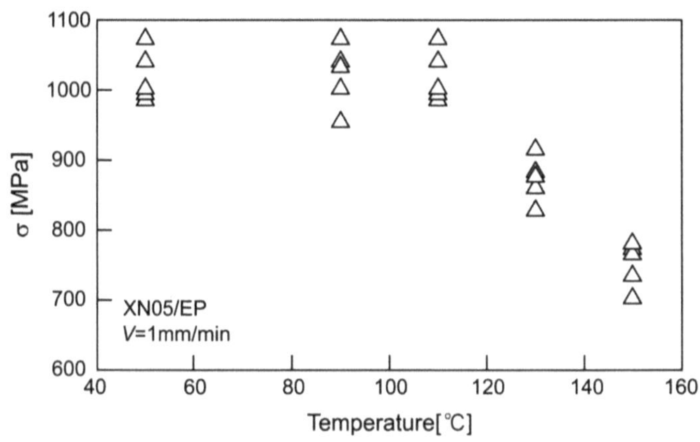

Fig. 13 Static strength versus temperature of XN05/EP

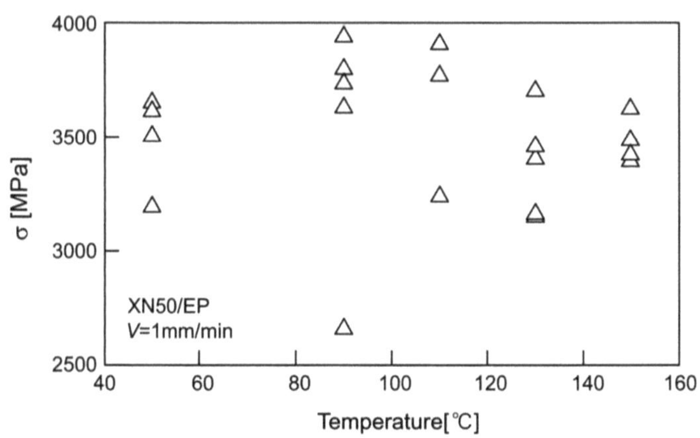

Fig. 14 Static strength versus temperature of XN50/EP

straight line with the slope of n_R, which is the viscoelastic parameter in Eqs. (1), (3), and (5). The viscoelastic parameters n_R of both CFRP strands are also completely different from each other.

Fractographs of both of XN05/EP and XN50/EP after the static tests show the same behavior as that for the case of PAN-based CFRP strands as shown in Fig. 17. The markedly uneven failure surface is observed in XN05/EP, where penetrations of cracks through fibers are not observed. A mirror surface with wide area is observed in XN50/EP, where the generated cracks penetrate through many fibers.

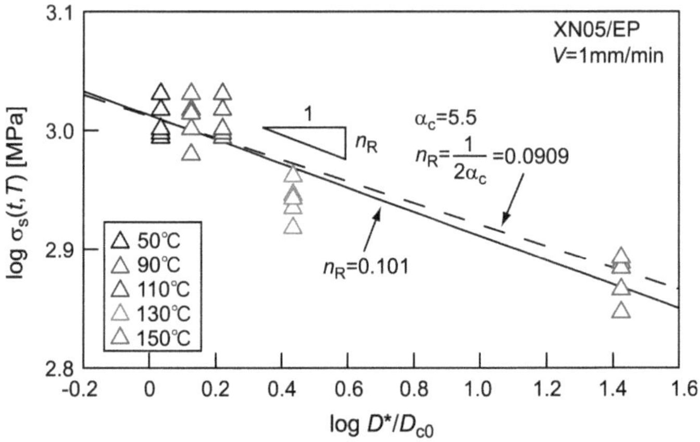

Fig. 15 Static strength of XN05/EP against viscoelastic compliance of matrix resin

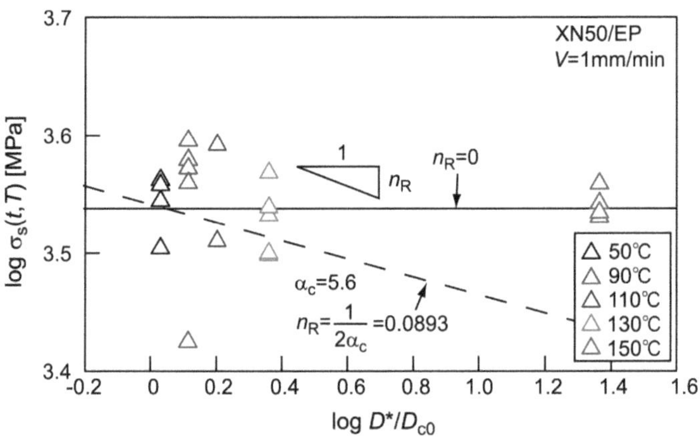

Fig. 16 Strength of XN50/EP against viscoelastic compliance of matrix resin

Fig. 17 Fractographs of XN05/EP and XN50/EP after static tension tests

8 Discussing of Failure Mechanisms

The experimental viscoelastic parameters n_R for T300/EP and XN05/EP as shown in Figs. 7 and 15 agree with the theoretical viscoelastic parameters n_R determined from $n_R = 1/2\alpha_c$ based on Rosen's shear lag model [11], where α_c is the shape parameter for tensile strength of single carbon fiber shown on Table 1, respectively. Therefore, it can be presumed that the failure mechanism of T300/EP and XN05/EP is based on Rosen's shear lag model. The failure of a carbon fiber based on Rosen's shear lag model is not influenced from other carbon fibers and therefore the markedly uneven failure surface should be observed on the fractograph of CFRP strand. Actually, the fractographs of T300/EP and XN05/EP show the uneven failure surfaces as shown in the left sides of Figs. 11, 12 and 17.

On the other hand, the failure of M40J/EP and XN50/EP show brittle fracture as shown in the right sides of Figs. 11, 12 and 17, and the static strengths of these CFRP strands do not show time and temperature dependent and the viscoelastic parameters n_R are nearly equal to zero. The reason why the statistical creep failure time of M40J/EP show clearly time dependent as shown in Fig. 10 is unclear.

9 Conclusions

We proposed a prediction method for statistical creep failure time under tension loading along the longitudinal direction of unidirectional CFRP using the statistical static tensile strength of CFRP strand and the viscoelasticity of matrix resin based on Christensen's model for viscoelastic crack kinetics. Results clarified that this prediction method is applicable for CFRP strands with high-strength PAN-based carbon fibers.

Statistical considerations related to strength should be associated strongly with the size effect of strength, as discussed in our recent paper [14]. Through these discussions, our results for resin-impregnated CFRP strands shall be extended in future studies to general CFRP structures that must have high reliability.

Statistical considerations related to the strength should be also associated with the water effect of strength for marine use. Our proposed methodology shall be useful for these discussions.

Acknowledgements The authors thank the Office of Naval Research for supporting this work through an ONR award to Dr. Yapa Rajapakse and Dr. Ming-Jen Pan. Our award, "Statistical Long Term Creep Failure Time of Unidirectional CFRP," is numbered N62909-16-1-2132. The authors thank Professor Richard Christensen of Stanford University as a partner on this project.

References

1. Aboudi J, Cederbaum G (1989) Analysis of viscoelastic laminated composite plates. Compos Struct 12:243–256
2. Sullivan J (1990) Creep and physical aging of composites. Compos Sci Technol 39:207–232
3. Gates T (1992) Experimental characterization of nonlinear, rate dependent behavior in advanced polymer matrix composites. Exp Mech 32:68–73
4. Miyano Y, Nakada M, McMurray MK, Muki R (1997) Prediction of flexural fatigue strength of CFRP composites under arbitrary frequency, stress ratio and temperature. J Compos Mater 31:619–638
5. Kawai M, Yagihashi Y, Hoshi H, Iwahori Y (2013) Anisomorphic constant fatigue life diagrams for quasi-isotropic woven fabric carbon/epoxy laminates under different hygro-thermal environments. Adv Compos Mater 22:79–98
6. Miyano Y, Nakada M, Cai H (2008) Formulation of long-term creep and fatigue strengths of polymer composites based on accelerated testing methodology. J Compos Mater 42: 1897–1919
7. Nakada M, Miyano Y (2015) Advanced accelerated testing methodology for long-term life prediction of CFRP laminates. J Compos Mater 49:163–175
8. Miyano Y, Nakada M, Kudoh H, Muki R (1999) Prediction of tensile fatigue life under temperature environment for unidirectional CFRP. Adv Compos Mater 8:235–246
9. Nakada M, Miyano Y, Kinoshita M, Koga R, Okuya T, Muki R (2002) Time-temperature dependence of tensile strength of unidirectional CFRP. J Compos Mater 36:2567–2581
10. Okuya T, Nakada M, Miyano Y (2013) Reliable test method for tensile strength in longitudinal direction of unidirectional carbon fiber-reinforced plastics. J Reinf Plast Compos 32:1579–1585
11. Nakada M, Okuya T, Miyano Y (2014) Statistical prediction of tensile creep failure time for unidirectional CFRP. Adv Compos Mater 23:451–460
12. Nakada M, Miyano Y (2016) Statistical creep failure time of unidirectional CFRP. Exp Mech 56:653–658
13. Christensen R, Miyano Y (2006) Stress intensity controlled kinetic crack growth and stress history dependent life prediction with statistical variability. Int J Fract 137:77–87
14. Christensen R, Miyano Y, Nakada M (2015) Size dependence of tensile strength for brittle isotropic materials and carbon fiber composite materials. Composites Science and Technology 106:9–14

Multiphysics Modeling of the Hygro-Mechanical Behavior of Heterogeneous Materials

Alexandre Clément, Sylvain Fréour and Frédéric Jacquemin

Abstract The present chapter investigates the modeling of the hygro-mechanical behavior of composites used in marine environment. The purpose of the proposed models is to analyze the effect of water absorption on a composite structure through mechanical quantities such as stress fields. The case of non-Fickian diffusion processes is first addressed with the Langmuir model allowing representing anomalies of diffusion. The elastic constitutive equations thus depend on the moisture content involving a hygroscopic strain field. The capabilities of the model are shown with a numerical study on a composite material. Second, we propose to introduce couplings between the water diffusion and the mechanical states. To achieve this, we mix the classical Fick model with the free volume theory and implement the model in a finite element software. A numerical study regarding the impact of such coupling on a pure resin case is carried out. We finally propose to take into account the uncertainties sometimes observed on the experimental data which enable the characterization of the diffusion properties. Some of the material properties of the hygro-elastic model thus become random and are modeled with random variables. The propagation of these uncertainties is made with a stochastic spectral approach based on polynomial chaos expansions. The efficiency of the proposed technique is shown with a numerical application involving a polyamide neat resin.

Keywords Composite materials · Aging · Transient hygro-mechanical problem · Multiphysics modeling · Uncertainties · Stochastic spectral approach

---

A. Clément (✉) · S. Fréour · F. Jacquemin
Institut de Recherche en Génie Civil et Mécanique, UMR CNRS 6183,
Université de Nantes, 58, rue Michel Ange, BP 420,
44606 Saint Nazaire, France
e-mail: alexandre.clement@univ-nantes.fr

P. Davies and Y.D.S. Rajapakse (eds.), *Durability of Composites in a Marine Environment 2*, Solid Mechanics and Its Applications 245,
https://doi.org/10.1007/978-3-319-65145-3_6

1 Introduction

Due to their excellent specific properties, the use of thermosetting and thermoplastic composites is growing in many industrial sectors such as transport and energy. Renewable marine energies are no exceptions since these materials favor the performances of energy converting systems [1]. However, marine structures are subjected to a harsh environment and moisture absorption may be harmful for the material. Most of polymeric resins employed in these composites are indeed hydrophilic and the water absorption may lead to a premature aging of the material. Different phenomena may occur and encourage damage processes. First, the absorbed water can react with the polymer network [2] and may decrease some mechanical properties such as the elastic modulus through plasticization for instance. Second, the moisture uptake yields a so-called hygroscopic swelling of the matrix and, since the fibers made of carbon or glass are hydrophobic, internal stresses occur within the composite which may deteriorate the structure [3]. It is therefore of first importance to be able to predict and analyze these phenomena in order to ensure the long-term behavior of the structure.

Many experimental studies have been devoted to water diffusion in composite materials such as the ones proposed in [4–6]. Most of them point out that the water uptake may be well represented with the classical uncoupled Fick model [7]. However, when a particular material exhibits deviations from this model, one should use more complex models to be able to represent the diffusion behavior. This task may be achieved using the dual stage Fick model [8] or the Langmuir model [9]. The latter appears as a good candidate since it may help to represent a wider range of sorption kinetics assuming that the absorbed moisture is composed of mobile and bound phases. However, unlike for the Fick model [10–12], numerical tools associated to the Langmuir model do not exist and transient local diffusion analysis is not possible for complex physical systems. The first part of this chapter is thus devoted to numerical transient uncoupled hygro-elastic analyses based on the Langmuir model for non-Fickian composite materials [13].

Even if the Langmuir model may improve the modeling of diffusion processes, the hygro-elastic problem remains uncoupled. However, a coupling between the mechanical fields and the water diffusion may exist in polymeric resins [14, 15] and composite materials [16–18]. For instance, a specimen subjected to tensile loading while aging tends to absorb water faster while a specimen subjected to compression tends to exhibit a lower water diffusivity. The second part of this chapter is devoted to the modeling of such couplings using the free volume theory which introduces a dependency of the local moisture field to the local strain field [19]. The nonlinear model has been implemented in a commercial finite element software allowing numerical analyses of this coupling.

The third and last part of this chapter deals with the handling of uncertainties appearing in water absorption phenomena. Experimental results may indeed exhibit relevant randomness on the diffusion properties of the studied polymeric resins and composites. Taking into account these uncertainties seems today essential if one seeks to obtain reliable numerical predictions, usable in a design process or decision-making. The most classical technique to achieve this kind of stochastic study relies on Monte Carlo simulations [20, 21]. This technique has been intensively used to solve complex stochastic physical problems. This nonintrusive approach only requires the use of a simple deterministic calculation code. However, it requires numerous deterministic computations which tend to lead to prohibitive computational costs. In order to bypass those issues, one can use the spectral stochastic methods proposed in [22] for the propagation of uncertainties through physical systems. These approaches provide a complete characterization of the random response under a suitable functional form and, once the solution is computed, allow a fast postprocessing of any probabilistic quantities such as statistics or probability laws. This parametric methodology, usually leading to construct polynomial chaos expansions of random variable or random fields, has been intensely developed to solve stochastic partial differential equations [23–28]. In this last chapter, we thus focus on the study of stochastic transient hygro-elastic behavior of composite materials based on a spectral stochastic approach.

2 Fickian and Non-Fickian Hygro-Mechanical Problem

In this section, we present the diffusion models used in this work. We first focus on the classical Fick model used in the following chapters for the study of couplings and uncertainties. We then present the Langmuir model allowing the modeling of moisture diffusion phenomena with anomalies of diffusion which cannot be represented with the Fick model. We also present the mechanical constitutive equations which take into account the hygroscopic swelling and help complete the hygro-elastic model. A numerical study on an epoxy-based composite reinforced with long carbon fibers is finally carried out.

2.1 *Microscale Problem*

We consider a heterogeneous material, schematically depicted in Fig. 1, which occupies a spatial domain $\Omega = \Omega_m \cup \Omega_r \in \mathbb{R}^d$ with $d \in \{1, 2, 3\}$. Ω_m and Ω_r, respectively, represent the matrix and the fibers acting as reinforcements. The variable $c(\mathbf{x}, t)$ denotes the moisture content of a material point, characterized by its position through vector $\mathbf{x}$, at time t. This moisture content $c(\mathbf{x}, t)$ is defined by

$$c(\mathbf{x}, t) = \frac{m_w(\mathbf{x}, t)}{m_0(\mathbf{x})}, \tag{1}$$

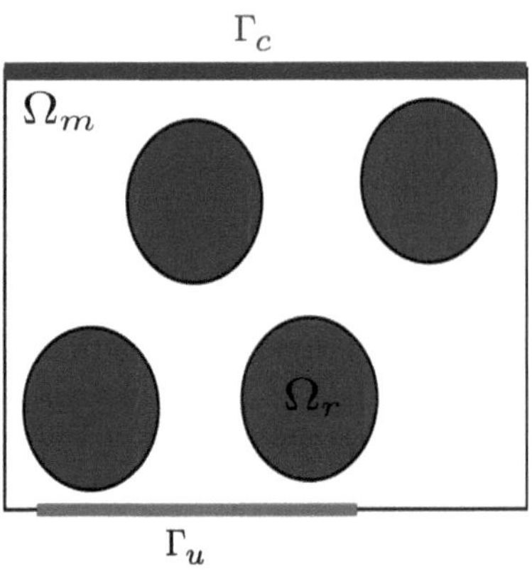

Fig. 1 Microscale model problem

where $m_w(\mathbf{x}, t)$ is the local uptake in mass of water whereas $m_0(\mathbf{x})$ is the local mass at the initial time.

The spatial average moisture content $\bar{c}(t)$ can be obtained with the following relation

$$\bar{c}(t) = \frac{1}{M_0} \int_{\Omega} \rho(\mathbf{x}) c(\mathbf{x}, t) d\Omega, \tag{2}$$

where $\rho(\mathbf{x})$ is the local density and M_0 is the mass of the sample at initial time. From an experimental point of view, using the knowledge of M_0 and the mass $M(t)$ at time t of the sample, the overall volume of which is Ω, $\bar{c}(t)$ is evaluated using the following relationship

$$\bar{c}(t) = \frac{M(t) - M_0}{M_0}. \tag{3}$$

Gravimetric measurements through time thus allow the characterization of the water diffusion kinetics. From now on, all equations will be written according to the local moisture content $c(\mathbf{x}, t)$. In this work, we assume that the diffusion process is governed by a unique diffusion coefficient D in each spatial direction. Moreover, since the fibers are considered hydrophobic, the problem may be solely formulated on the domain Ω_m. This remark is true for both models which are exposed in the following.

The Fick law is a classical choice to represent a diffusion process where each water molecule is free to move in the polymer network associated with domain Ω_m. Since the diffusivity is assumed constant (i.e., independent of moisture content or mechanical states), the Fick local diffusion problem can be posed as: find the solution field $c(\mathbf{x}, t)$ such that it verifies

$$\begin{aligned} \frac{\partial c}{\partial t} &= D\Delta c \quad on \ \Omega_m, \\ c &= c_{imp} \quad on \ \Gamma_c, \end{aligned} \tag{4}$$

where c_{imp} is a given moisture content applied on a part Γ_c of boundary $\partial\Omega$.

We now focus on the Langmuir diffusion model presented in [9], where the authors divided the water molecules into two populations. The first $n(\mathbf{x}, t)$ is free to move while the molecules of the second phase $N(\mathbf{x}, t)$ are bonded to the polymer network due to reversible chemical reactions such as those occurring when moisture induced plasticization takes place. A free water molecule could become a bonded one with a frequency α and reciprocally a bonded molecule could be freed with a frequency β. The total moisture content $c(\mathbf{x}, t)$ naturally verifies the expression:

$$c(\mathbf{x}, t) = n(\mathbf{x}, t) + N(\mathbf{x}, t). \tag{5}$$

The Langmuir local diffusion problem can be written as: find the solution fields $n(\mathbf{x}, t)$ and $N(\mathbf{x}, t)$ such that they verify

$$\begin{aligned} \frac{\partial n}{\partial t} + \frac{\partial N}{\partial t} &= D\Delta n && \text{on } \Omega_m, \\ \frac{\partial N}{\partial t} &= \alpha n - \beta N && \text{on } \Omega_m, \\ n &= n_{imp} && \text{on } \Gamma_c, \\ N &= N_{imp} && \text{on } \Gamma_c, \end{aligned} \tag{6}$$

where n_{imp} and N_{imp} are, respectively, the imposed free and bounded moisture content on Γ_c. Considering those two different populations, the Langmuir model is able to represent a wider class of diffusion phenomena than the Fick model. In particular, delay or, on the contrary, fast absorption, at early instants of diffusion can be simulated with this model.

Both previous diffusion problems can be coupled to an elastic problem in order to analyze the deformation of the material or the structure under aging conditions. We denote by $\mathbf{u}(\mathbf{x}, t)$ the displacement field, by $\boldsymbol{\varepsilon}(\mathbf{u}(\mathbf{x}, t))$ the strain tensor, and by $\boldsymbol{\sigma}(\mathbf{x}, t)$ the Cauchy stress tensor. Both the reinforcements and the matrix are assumed to be linear isotropic elastic materials represented by the fourth order stiffness tensor $\mathbf{C}$ verifying

$$\mathbf{C}(\mathbf{x}) = \begin{cases} \mathbf{C}_m & \text{if } \mathbf{x} \in \Omega_m \\ \mathbf{C}_r & \text{if } \mathbf{x} \in \Omega_r \end{cases}, \tag{7}$$

where $\mathbf{C}_m$ and $\mathbf{C}_r$ are constant tensors. Moreover, β_h is the hygroscopic expansion coefficient which is considered, in this work, identical in each direction. The hygroscopic expansion is represented by the diagonal tensor $\boldsymbol{\beta}_h$ whose diagonal components are equal to β_h. Material parameter β_h is taken equal to 0 for the hydrophobic reinforcements. Finally, the quasi-static linear elastic problem can be written as: find the displacement field $\mathbf{u}(\mathbf{x}, t)$ such that

$$\begin{aligned} \nabla \cdot \boldsymbol{\sigma} &= 0 && on\ \Omega, \\ \boldsymbol{\sigma} &= \mathbf{C} : [\boldsymbol{\varepsilon} - \boldsymbol{\beta}_h c] && on\ \Omega, \\ \mathbf{u} &= \mathbf{u}_{imp} && on\ \Gamma_u, \end{aligned} \tag{8}$$

where $\mathbf{u}_{imp}$ is the imposed displacement on the part Γ_u of $\partial\Omega$ and $c(\mathbf{x}, t)$ is the solution field of either problem (4) or problem (6).

In this work, simulations involving Fick or Langmuir's models were performed using the commercial finite element software Abaqus™ in which a specific element was developed using an UEL subroutine. The corresponding elastic problems have also been solved using the same finite element software. The next section is devoted to a numerical study involving a composite material with a non-Fickian diffusion behavior.

2.2 *Numerical Study: Composite Material with a Non-Fickian Diffusion Behavior*

We focus on an epoxy-based composite reinforced with carbon fibers and propose to analyze its diffusion and elastic behavior during aging. The gravimetric measurements characterizing the diffusion process are coming from [8]. The diffusion parameters corresponding to Fick and Langmuir models can be determined from a minimization problem aiming at finding the optimal diffusion parameters reducing the least-square distance between the experimental data and the analytical solution given in [7] for the Fick model and in [9] for the Langmuir model. Figure 2 illustrates the experimental sorption curve for the epoxy resin with the results of the

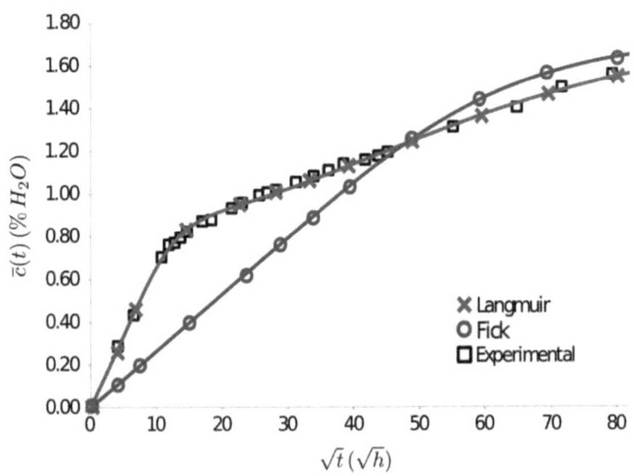

Fig. 2 Experimental diffusion data of the studied epoxy resin with both Fick and Langmuir identification results

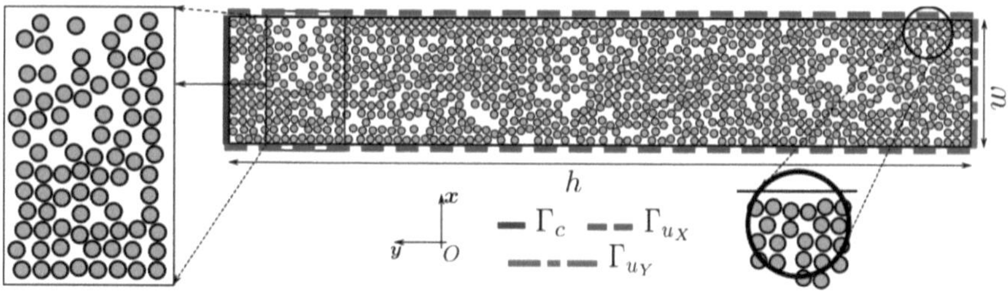

Fig. 3 Problem definition with the different types of studied microstructures

identification for the Fick and the Langmuir models. We can notice that this resin presents an anomalous diffusion behavior and only the Langmuir model allows a good representation of the diffusion process. In the following, only the Langmuir model will thus be used. The Langmuir identification leads to the following results: $D = 2.04\,\mu m^2 s^{-1}$, $C^\infty = 1.71\%$, $\alpha = 0.717\ 10^{-7} s^{-1}$ and $\beta = 0.745\ 10^{-7} s^{-1}$.

Figure 3 illustrates the geometry of the problem. We focus on a square sample of size 75 mm where thickness e is sufficiently small to neglect the water absorption by the edges. Only a slice of the superior half of the sample is therefore geometrically modeled. The geometrical characteristics are the following: the height h = 600 μm and the width w = 100 μm. We propose to perform the same study with different microstructures for which the spatial distribution of the fibers can be periodic or random. One should note that only a random configuration is illustrated in Fig. 3. The diameter of fibers is equal to 7 μm and the volume fraction is equal to 61%. The elastic and other material parameters are the following ones for the epoxy resin: Young's modulus $E_m = 4$ GPa, Poisson's ratio $\nu_m = 0.36$, density $\rho_m = 1310\,\text{kg}\,\text{m}^{-3}$, and hygroscopic coefficient $\beta_h^m = 0.00324$. For the carbon reinforcements, the corresponding parameters are: Young's modulus $E_f = 20$ GPa, Poisson's ratio $\nu_f = 0.33$, density $\rho_f = 1780\,\text{kg}\,\text{m}^{-3}$, and hygroscopic coefficient $\beta_h^f = 0$.

Figure 4 presents the sorption curve for the different geometries of the microstructure (one is regular and four are random). The sorption curves are very similar even for the periodic microstructure. Only small deviations may be seen at the beginning of the process where the diffusion is faster in the periodic case. Fibers clusters which occur in random cases indeed act as barriers. Moreover, it may be noticed that the sorption curves predicted for the composites structure are typical of an anomalous diffusion behavior. Furthermore, their shapes are similar to the sorption curve characterizing the pure resin sample (cf. Fig. 2).

We then focus on the elastic results. Figure 5 illustrates the local stress fields σ_{yy} after 1 h and when the steady state is reached. We can observe that the extreme stresses are localized in the matrix and that material points surrounded by fibers are in traction while the other points located in a zone with a lack of reinforcements are under compression. Moreover, this figure highlights the geometrical impact on stress intensity since random cases lead to higher local stresses regarding the

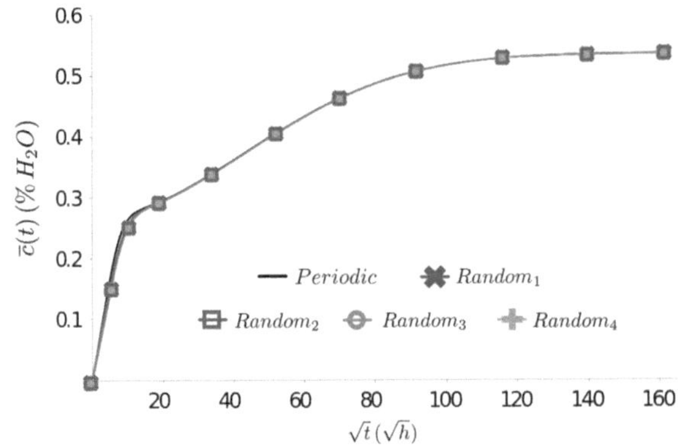

Fig. 4 Average water content $\bar{c}(t)$ with respect to time t for periodic and random geometries

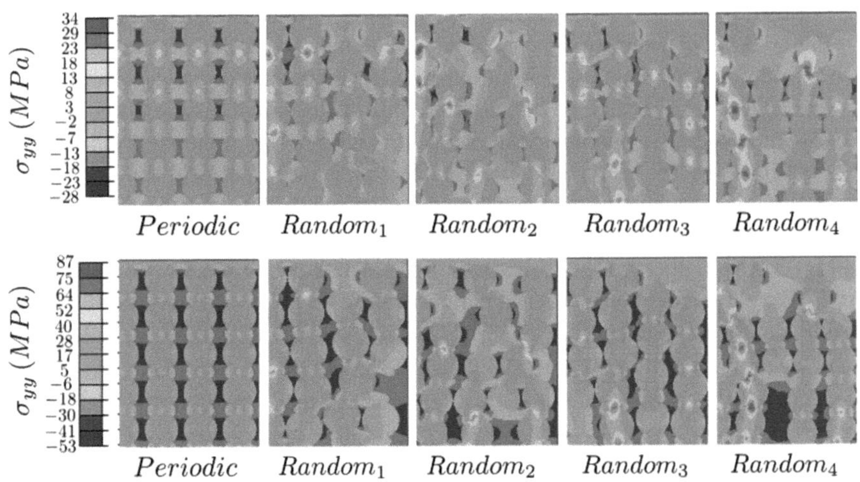

Fig. 5 Local stress fields σ_{yy} at 1 h (*up*) and when the steady state is reached (*down*)

periodic case. If the average water content is not modified by the geometrical distribution of the fibers, local quantities such as local stress fields strongly depend on this distribution. This assessment is emphasized with Fig. 6a which illustrates the spatial distribution of the local stresses at steady state. On this figure, the local stresses within the matrix and the reinforcements are separated in order to clearly observe the differences between the two constituents.

We finally focus on spatial average quantities which can be seen as macroscopic results. Figure 6 also presents the spatial mean stress $\overline{\sigma_{yy}}$ within the matrix and the reinforcements. One should note that results for random or periodic configurations lead to the same values, this is why only one configuration is illustrated in this figure. It shows that the matrix is globally under compression while the fibers are in

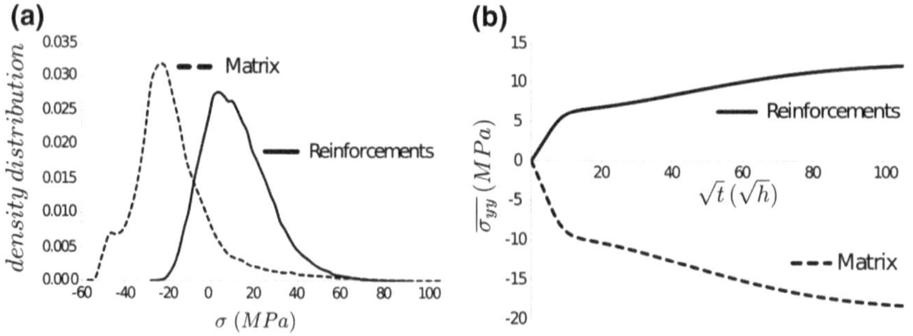

Fig. 6 Spatial density distribution of the local stresses σ_{yy} at the steady state (**a**) and spatial average stress $\overline{\sigma_{yy}}$ within the matrix and the reinforcements (**b**)

tension. However, compared to observations made on Fig. 5, it highlights the loss of information resulting from considering the spatial mean values. Indeed, the matrix or the fibers could locally be in tension while the mean stress is negative or in compression while the mean stress is positive. In conclusion, changes in the distribution of the fibers only affect the local fields; the global fields remains almost the same.

This numerical study pointed out the use of the improved Langmuir diffusion model in order to perform a numerical investigation of the humid aging of composite materials. In addition, the geometrical effect on local field, particularly for the stresses induced by the swelling of the matrix, between regular and random geometries has been highlighted. However, in the section, we only consider an uncoupled hygro-elastic problem since the diffusion or mechanical properties do not vary according to the various local fields. The purpose of the following section is to introduce such dependencies between the diffusion coefficient and the mechanical fields.

3 Coupled Hygro-Elastic Problem

The Fickean model assuming a constant diffusivity is widely used in numerical computations [3, 12, 29]. However, experimental work has emphasized that the diffusivity can depend on the mechanical field, for both neat resin and composite samples [14–17, 30]. This coupling between the diffusion kinetics and the mechanical states has been highlighted through aging tests under mechanical loads. They emphasize a nonlinear relationship between the applied stresses and the global (or macroscopic) diffusivity of the studied samples. In this work, such dependency is modeled using the free volume theory briefly presented in the following. A numerical example allows the results to be compared between the coupled hygro-elastic model and the classical uncoupled Fick model and emphasizes the impact of such couplings on the hygro-mechanical behavior.

3.1 Free Volume Theory

Recently, the coupling between the diffusion coefficient and the mechanical field was expressed not using the stress tensor $\boldsymbol{\sigma}$ but the strain $\boldsymbol{\varepsilon}$ [19]. This choice implies that the mobility of water molecules, in the polymer network, is physically related to the space available embodied by the free volume V_f. This volume, in blue on Fig. 7, is defined as the difference between the macroscopic observable volume V and the real volume occupied by the polymer molecules V_p (colored in red on Fig. 7). On this figure, the black border denotes the outer surface of the sample, the red circular areas are the regions occupied by the polymer, whereas the blue areas denote the free volume existing between the polymer molecules.

We thus propose to relate the diffusivity to the free volume by the Doolittle equation as has been done in [14]. The ratio between the diffusivity $D(\boldsymbol{\varepsilon}, t)$ of the polymer in a given strain state and the corresponding reference value in the strain free state D_0 is given by the following equation:

$$ln\left(\frac{D(\boldsymbol{\varepsilon}, t)}{D_0}\right) = a\left(\frac{1}{v_0^f} - \frac{1}{v^f(\boldsymbol{\varepsilon}, t)}\right), \tag{9}$$

where v_0^f and $v^f(\boldsymbol{\varepsilon}, t)$ are, respectively, the free volume fractions of the strain free and strained polymer and parameter a is empirical. In addition, v_0^f and $v^f(\boldsymbol{\varepsilon}, t)$ are related through:

$$v^f(\boldsymbol{\varepsilon}, t) = v_0^f + \frac{\Delta V(\boldsymbol{\varepsilon}, t)}{V_0}, \tag{10}$$

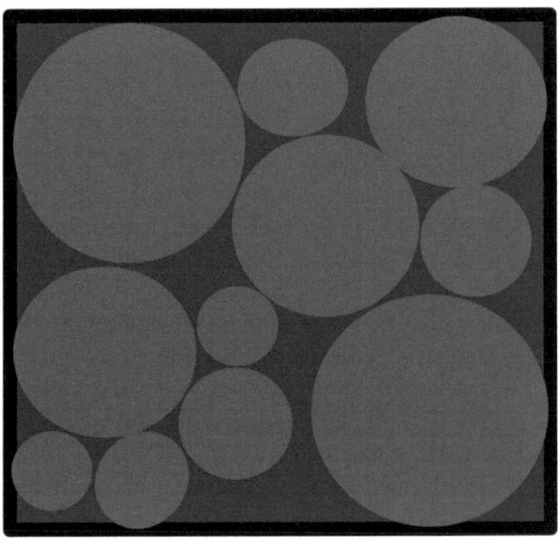

Fig. 7 Definition of the free volume in a neat resin specimen

where $\Delta V(\boldsymbol{\varepsilon}, t)$ is the volume variation induced by the mechanical loading and V_0 stands for the volume in a free of stress/strain state. Assuming small displacements, the ratio $\frac{\Delta V(\boldsymbol{\varepsilon}, t)}{V_0}$ may be approximated by the trace of the strain tensor denoted by $tr\, \boldsymbol{\varepsilon}$. Equation (9) may thus be written:

$$D(\boldsymbol{\varepsilon}) = D_0\, exp\left(\frac{a\, tr\, \boldsymbol{\varepsilon}}{v_0^f (v_0^f + tr\, \boldsymbol{\varepsilon})}\right). \tag{11}$$

Using Eq. (11), the Fick local diffusion problem becomes: find the solution field $c(\mathbf{x}, t)$ such that:

$$\begin{aligned} \frac{\partial c}{\partial t} &= \nabla \cdot (D(\boldsymbol{\varepsilon}, t)\nabla c) \quad \text{on } \Omega_m \times T, \\ c &= c_{imp} \quad \text{on } \Gamma_c \times T. \end{aligned} \tag{12}$$

In fact, the maximal moisture absorption capacity also depends on the local strain [14]. However, this second coupling is more difficult to take into account in the finite element framework and is not considered in this work. As for the Langmuir model, dedicated Abaqus™ subroutines have been implemented in order to solve the nonlinear problem resulting from the coupled hygro-elastic model.

Considering experimental data such as gravimetric measurements, the difficulty comes from the identification of the free volume theory parameters for which no analytical solutions are available. Different authors seem to agree for pure resin without defect about the value of $v_0^f = 2.5\%$ [14, 18]. Moreover, tests results given in [30] show that the parameter $a \in [0.031;\ 0.036]$ for epoxy resins. Therefore, both a and v_0^f are fixed in this work and only D_0 remains unknown. This latter parameter is determined through an inverse problem consisting of finding the optimal value of D_0 using finite element simulations and experimental data.

3.2 *Numerical Study: Humid Aging Under Three Point Bending Load*

In this section, we propose to analyze the aging of an epoxy resin submitted to a three point flexural bending. The test is depicted in Fig. 8 with its diffusion and mechanical boundary conditions. Two values for the external force F are considered: $F_1 = 5\,\text{N}$ and $F_2 = 10\,\text{N}$. The coupons are 80 mm long and 3 mm thick. The distance between the supports is $d = 48$ mm. The diffusion parameters of the epoxy resin are $D_0 = 0.696\,\mu\text{m}^2\text{s}^{-1}$, $a = 0.031$, $v_0^f = 2.5\%$ and $C^\infty = 3\%$. The elastic constants for the same material are $E = 2.7$ GPa, $\nu = 0.35$ and hygroscopic coefficient $\beta_h = 0.0025$.

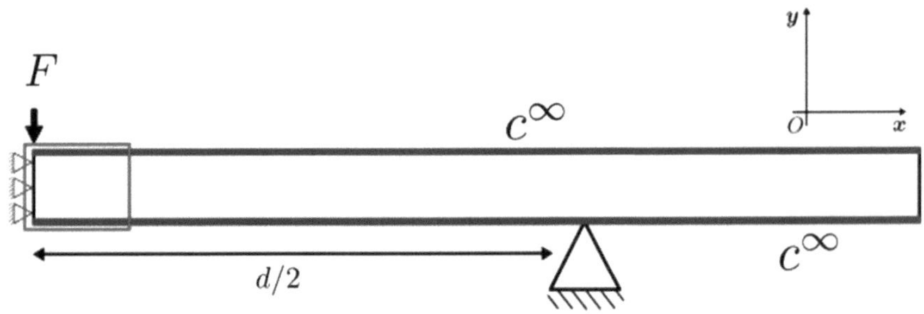

Fig. 8 Description of the flexural three points bending test with imposed boundary conditions

We first compare the numerical results for coupled and uncoupled Fick models. The moisture content fields $c(\mathbf{x})$, obtained from computations with both uncoupled and coupled models, at time t = 144,000 s, close to the middle of the specimen, are shown in Fig. 9. It is worth noting that for the lower force, the field is almost symmetrical as in the uncoupled case. This is no longer true for $F_2 = 10\,\mathrm{N}$ where the field is strongly disturbed in the area under the applied force F_2. Figure 2b presents the gradients of $c(\mathbf{x})$ through the thickness at two specific times t = 72,000 s and t = 720,000 s, in the middle section of the specimen. At t = 72,000 s, all coupled models lead to unsymmetrical results unlike the Fick model. Close to

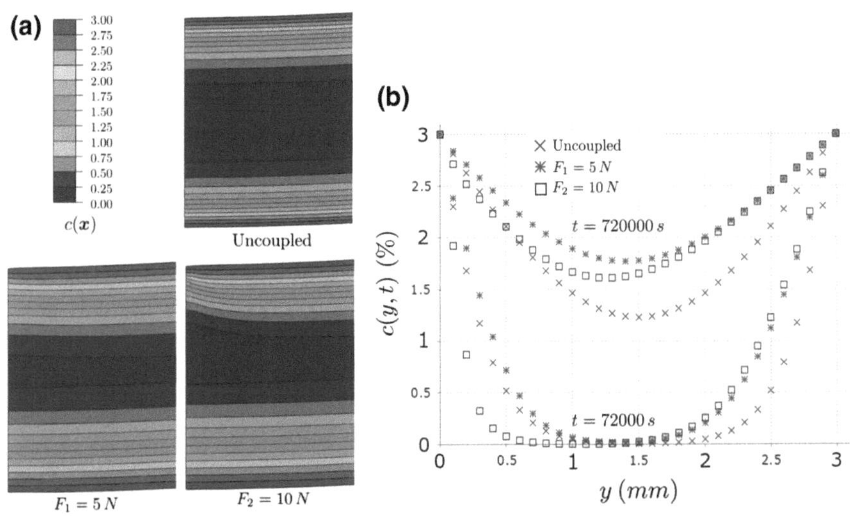

Fig. 9 Moisture field $c(\mathbf{x})$ under a three point flexural bending load F for both uncoupled and coupled problems (**a**). Moisture gradient through the thickness in the middle of the specimen for the uncoupled problem, coupled with F_1 and coupled with F_2 at two transient time steps t = 72000 s and t = 720000 s (**b**)

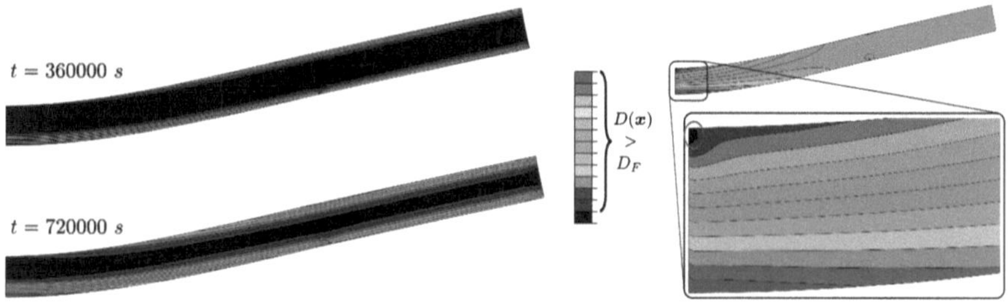

Fig. 10 Diffusivity field $D(\mathbf{x}, t)$ at two transient time steps (**a**) and at saturation (**b**) in the case $F_2 = 10\,\mathrm{N}$

the saturation, the coupled model tends to lead to more symmetrical results. As expected, on the top face, the absorption is slowed down, while the diffusivity is higher on the bottom side of the coupon. This is understandable, because the material is in a compression state close to the loaded surface and in tension on the other side.

The local diffusivity $D(\mathbf{x}, t)$, at t = 360,000 s, t = 720,000 s and at the steady state, for $F_2 = 10\,\mathrm{N}$, is plotted on Fig. 10. The area where $D(\mathbf{x}, t) > D_F$ are colored. According to Fig. 10, at t = 360,000 s, $D(\mathbf{x}, t)$ was predicted to remain lower than D_F in the largest part of the specimen. Furthermore, the asymmetry appeared clearly. At t = 720,000 s, $D(\mathbf{x}, t)$ remains lower only in a tiny area under the application point of F. The impact of both model and force intensity on local fields is thus emphasized. Those observations may be extrapolated to larger dimension immersed structures such as tidal blades. In such structures, due to external loads, moisture fields are heterogeneous, generating internal stresses which may lead to local damage and a premature deterioration.

We now focus on the predictions of the evolution of the global water content. Figure 11 illustrates the numerical sorption curves obtained with the uncoupled Fick model and the coupled model for forces F_1 and F_2. We can notice that all curves are very close to each other. This assessment is similar to the one made in the previous section dedicated to uncoupled models. Even if local fields are significantly affected by the couplings, the global quantity tends to remain constant. Thus, even when strong couplings occur at the microscopic scale, global experimental data may still follow the uncoupled Fick law. However, the reader must keep in mind that the maximum water content is assumed constant in this work; introducing a coupling between this diffusion parameter and the mechanical strain field may change this statement. Experimental data on such mechanical tests performing during aging should be made in the future to confirm, or not, those conclusions.

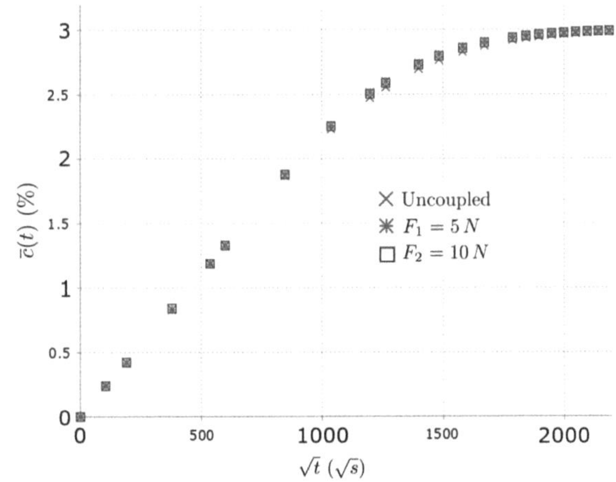

Fig. 11 Evolution of C with respect to $\sqrt{t}$ for the uncoupled case and the coupled cases associated to F_1 and to F_2

4 Stochastic Hygro-Elastic Problem

This last section is devoted to taking into consideration the uncertainties which may appear in a hygro-elastic problem. Figure 12 presents the moisture uptake through time of polyamide composite samples reinforced with glass fibers. Even if all samples are assumed to be the same (i.e, same material and same geometry), we can observe significant differences for both coefficients of diffusion and maximum water content. Taking into account solely the mean value or extreme values of the diffusion parameters may lead to quite poor predictions. Probabilistic approaches thus aim at considering the uncertainties observed on the input data and at propagating them in the physical system. The response thus becomes random and allows more robust and reliable predictions. The following section features a methodology to efficiently propagate the uncertainties based on a parametric vision and

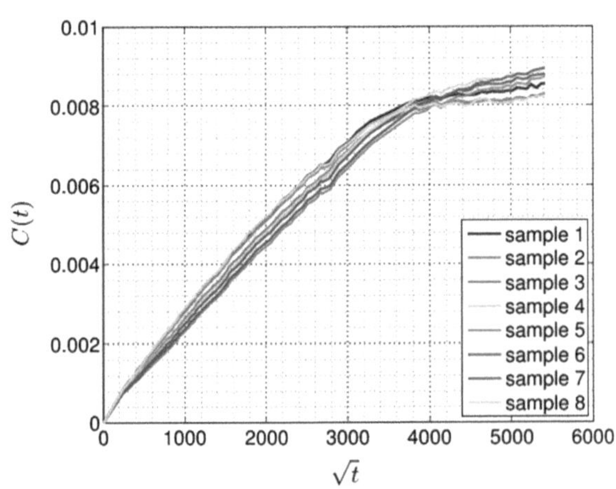

Fig. 12 Experimental absorption curves for a polyamide/glass unidirectional composite

polynomial chaos approximation. A stochastic numerical study illustrates the capabilities of the proposed technique.

4.1 Stochastic Model and Polynomial Chaos Approximation

We are now interested in taking into account the uncertainties on the diffusive properties of the material. For the sake of simplicity, we solely present the stochastic problem for the diffusion problem (4) corresponding to the Fick model. To achieve this task, we adopt a parametric vision of the uncertainties resulting in working in a finite dimension probability space. The probabilistic content is then represented by a set of independent random variables $\boldsymbol{\xi} = (\xi_1, \ldots, \xi_m)$ with probability law $P_{\boldsymbol{\xi}}$. We introduce the associated probability space $(\boldsymbol{\Theta}, B_{\Theta}, P_{\xi})$, where $\boldsymbol{\Theta} \subset \mathbb{R}^{\mathrm{m}}$ is the set of elementary events and B_{Θ} is a σ-algebra on $\boldsymbol{\Theta}$. Random variables $\boldsymbol{\xi}$ are used to characterize the various random inputs, such as the material properties or loadings.

The introduction of uncertainties modifies problem (4) which becomes: find local moisture field $c(\mathbf{x}, t, \xi) : \Omega_m \times \mathrm{T} \times \boldsymbol{\Theta}$ such that

$$\begin{aligned} \frac{\partial c(\mathbf{x}, t, \boldsymbol{\xi})}{\partial t} &= D(\boldsymbol{\xi}) \Delta c(\mathbf{x}, t, \boldsymbol{\xi}) \quad \text{on } \Omega_m \times T \times \boldsymbol{\Theta}, \\ c(\mathbf{x}, \boldsymbol{\xi}) &= c_{imp}(\mathbf{x}, \boldsymbol{\xi}) \qquad\quad \text{on } \Gamma_c \times \boldsymbol{\Theta}, \end{aligned} \tag{13}$$

In order to quantify the random local response $c(\mathbf{x}, t, \xi)$, it is necessary to use dedicated computational techniques. The best known is the Monte Carlo method which consists of generating a large number of simulations of $\boldsymbol{\xi}$ and evaluating the system response $c(\mathbf{x}, t, \xi)$ for each of them. This nonintrusive technique is easy to implement and especially provides a good estimate of the statistical moments. However, it may require a large number of deterministic calculations and leads to significant computational times. We therefore choose to use a different calculation technique here, belonging to spectral stochastic methods, and leading to a functional representation of the random response, which requires a few deterministic calculations when the number of random input variables m is low enough.

As has been done for the spatial time problem for which we introduced a spatial approximation space thanks to finite elements, it is necessary to define a stochastic approximation space $\mathsf{S}_P \subset \mathsf{S} = L^2(\boldsymbol{\Theta}, dP_{\boldsymbol{\xi}})$ where S is the second-order random variable space. The stochastic approximation space S_P can always be written;

$$\mathsf{S}_P = \{v(\xi) = \sum_{\alpha=1}^{P} v_\alpha H_\alpha(\boldsymbol{\xi}), v_\alpha \in \mathsf{R}\}, \tag{14}$$

where $\{H_\alpha\}_{\alpha=1}^{P}$ is a functional basis of S for which several choices exist such as polynomial chaos expansion and its generalization [22, 31–33]. These types of representation are well adapted for regular functions with respect to input random variables. When those functions are less regular (for instance a continuous function whose derivatives are discontinuous) a high number P of basis functions may be needed. In this case, other ways of representation may be used such as piecewise polynomial approximation [34, 35] or multi-wavelets [25, 36].

In order to solve the problem (13), we choose to use a L^2 projection method at the stochastic level [37–39] which only requires the resolution of problem (4) for a final set of realizations of basic random variables.

The aim is to seek a functional form of the semi-discretized solution $\mathbf{c}(t_n, \xi) \in \mathrm{R}^Q \forall 0 \leq t_n \leq T$ where $Q = N \times P$ and such that:

$$\mathbf{c}(t_n, \boldsymbol{\xi}) \approx \sum_{\alpha=1}^{P} \mathbf{c}_\alpha(t_n) H_\alpha(\boldsymbol{\xi}) \tag{15}$$

The L^2 projection technique consists in defining approximation (15) as the projection of $\mathbf{c}(t_n, \boldsymbol{\xi})$ onto the subspace S spanned by the basis functions $\{H_\alpha\}_{\alpha=1}^{P}$. This projection is defined with respect to the usual inner product of S

$$\langle v, w \rangle_{\mathsf{S}} = E(v(\boldsymbol{\xi}) w(\boldsymbol{\xi}))) = \int_{\boldsymbol{\Theta}} v(\mathbf{y}) w(\mathbf{y}) dP_\xi(\mathbf{y}). \tag{16}$$

When considering orthonormal functions H_α, the coefficients $\mathbf{c}_\alpha(t_n) \in \mathsf{R}^N$ of the approximate solution (15) are then defined by

$$\mathbf{c}_\alpha(t_n) = E(\mathbf{c}(t_n, \boldsymbol{\xi}) H_\alpha(\boldsymbol{\xi})), \tag{17}$$

where $E(\cdot)$ denotes the mathematical expectation. Computing coefficients $\mathbf{c}_\alpha(t_n)$, and thus the discretized solution leads to an approximation of the expectation from below. This task can be achieved with a suitable numerical integration at the stochastic level (quadrature, Monte Carlo, etc.) which is written as:

$$\mathbf{c}_\alpha(t_n) = \sum_{k=1}^{K} \omega_k \mathbf{c}(t_n, \boldsymbol{\xi}_k) H_\alpha(\boldsymbol{\xi}_k), \tag{18}$$

where the $\boldsymbol{\xi}_k$ and ω_k are, respectively, the integration points and the associated integration weights. These points correspond to particular elementary events of input random variables ξ. This approach is thus clearly a nonintrusive technique since it only requires the resolution of K deterministic problems, for instance, with the use of a simple deterministic finite element solver. In the following, the proposed approach will be referred to as S-FEM for stochastic finite elements method. In the next section, we propose a numerical study showing the capabilities of the proposed S-FEM approach.

4.2 Numerical Example: Stochastic Hygro-Elastic Study of a Polyamide Neat Resin

The numerical study deals with a PA6 neat resin. We focus on a 2D representation of experimental samples with small thickness leading to the problem of a plate of size $1 \times 10\,\text{mm}^2$ thanks to symmetric conditions. The probabilistic content is represented by two independent Gaussian random variables characterized by their means and standard deviations which model the isotropic diffusion coefficient and the maximum moisture content such that $D \in \mathrm{N}(1.19\,10^{-7},\, 1.2\,10^{-8})$ (mm^2/s) and $C_\infty \in \mathrm{N}(3.81, 0.06)$ (%). The statistic information on these two random variables comes from experimental observations of 10 different samples of the same material. For each of them, an identification procedure has allowed the Fick diffusion parameters to be determined. We can notice that the scatter is more important for the diffusion coefficient than for the maximum moisture content. The elastic parameters are all deterministic with Young's modulus $E = 3.8$ GPa, Poisson's ratio $\nu = 0.35$, and isotropic hygroscopic expansion coefficient $\beta = 0.00295$. The initial water content is assumed equal to zero within the sample. For the spatial approximation, we use a finite element mesh composed of 8000 3-node linear triangle elements. For the time approximation, an implicit Euler forward scheme is used with a time step increment of 2 days. At the stochastic level, we use a polynomial chaos with degree $p = 3$, and basis function $\{H_\alpha\}$ are the Hermite polynomials. Note that both spatial, time and stochastic convergence have been checked to provide accurate results in the following.

We first focus on the global moisture content. Figure 13 illustrates the 99% confidence interval (CI) and the mean of the stochastic process $C(t)$ obtained with the S-FEM solution. On the same figure, the experimental data are depicted showing a good agreement with the stochastic numerical predictions. The proposed

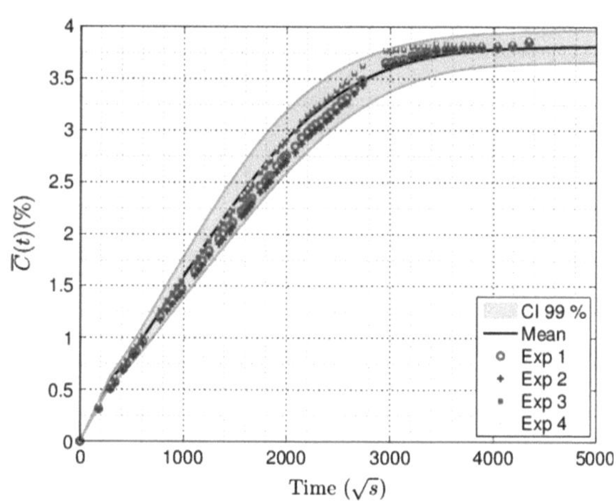

Fig. 13 Evolution of the stochastic global water content $C(t)$ with respect to time; comparison between the S-FEM solution and the experimental data

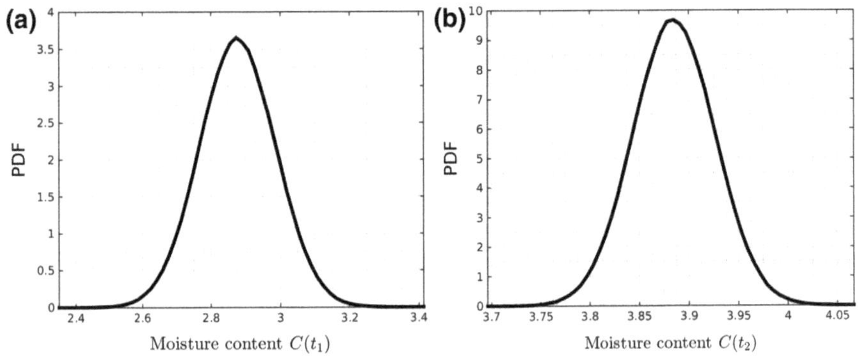

Fig. 14 Probability density functions of moisture content $C(t)$ for two different time steps $t_1 =$ 46 days and $t_2 = 185$ days

technique thus allows reliable numerical predictions regarding the stochastic diffusion process.

The proposed method allows a fast postprocessing of statistical or probabilistic quantities such as density probability functions needed in the computation of probabilities of particular events. Figure 14 presents the probability density functions (PDF) of stochastic process $C(t)$ for two different time steps $t_1 = 46$ days and $t_2 = 185$ days. We can observe that both PDF tend to be very close to a Gaussian distribution which characterizes the input random variables.

Owing to the diffusion model, we can have access to the uncertainties on local quantities for which no experimental data are available. The evolution of the stochastic local water content for two different material points, $\mathbf{x}_1$, located close to the boundary, and $\mathbf{x}_2$, located in the middle of the plate, is presented on Fig. 15. We can observe that the moisture uptake starts earlier for the point $\mathbf{x}_1$ since it is closer to the boundary. Moreover, we notice that the level of uncertainties is a little higher

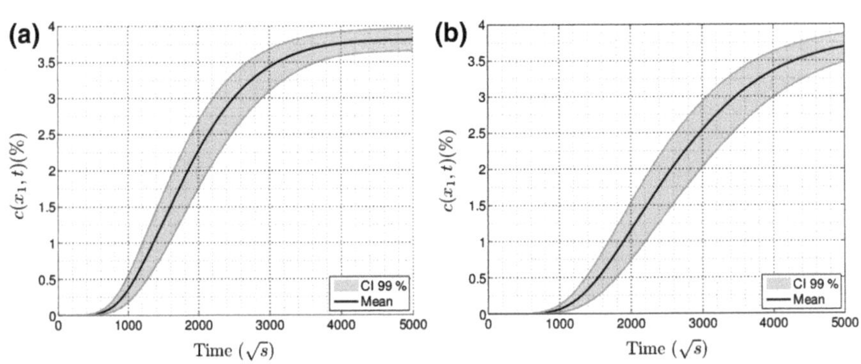

Fig. 15 Evolution of the stochastic local water content $c(\mathbf{x}, t)$ with respect to time for two different material points $\mathbf{x}_1$, close to the boundary (**a**), and $\mathbf{x}_2$ in the middle of the plate (**b**)

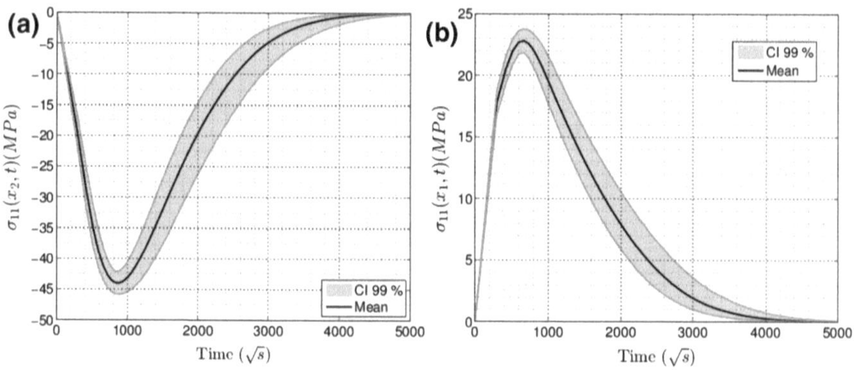

Fig. 16 Evolution of stochastic process $\sigma_{xx}(\mathbf{x}, t)$ with respect to time for two different material points $\mathbf{x}_1$ (**a**) and $\mathbf{x}_2$ (**b**)

for point $\mathbf{x}_2$ than for point $\mathbf{x}_1$. Indeed, since the scattering for the maximum water content acting as the boundary condition is lower compared to the diffusion coefficient, material points located close to this boundary are less affected. However, points located quite far from this boundary, such as $\mathbf{x}_2$, are more sensitive to the uncertainties on the diffusion coefficient.

We finally focus on elastic quantities through the stress field $\sigma_{xx}(\mathbf{x}, t)$. We analyze the uncertainties on this quantity for material points $\mathbf{x}_1$ and $\mathbf{x}_2$. For both points, we can observe relevant level of stresses at the beginning of the diffusion process. Moreover, we can notice that point $\mathbf{x}_1$ is in a compression state while $\mathbf{x}_2$ is in a tension state during the whole duration of the diffusion process. However, for both points, the stresses tend to zero when the steady state is reached since the water content field and then, the hygroscopic strain field, are homogeneous within the material. Finally, we can observe that uncertainties are also significant for the elastic quantities with levels similar to the ones noted for the moisture content (Fig. 16).

Figure 17 presents the evolution of the stochastic process $u_x(\mathbf{x}, t)$ for a material point located between the boundary and the center of the plate. We can observe that the displacement increases quite fast and tends to be quite similar to the evolution of the local moisture uptake with uncertainties of the same range. On the same figure is depicted the density probability function of the same local quantity at the steady state. We note that this PDF (or the corresponding CDF) may be used to evaluate failure probability; for instance, if a specific displacement criterion implies that the local displacement should not exceed 5.8 10^{-3} mm we have the following probability $P(u_x^{\max} > 5.8\,10^{-3}) = 5.5\ 10^{-3}$.

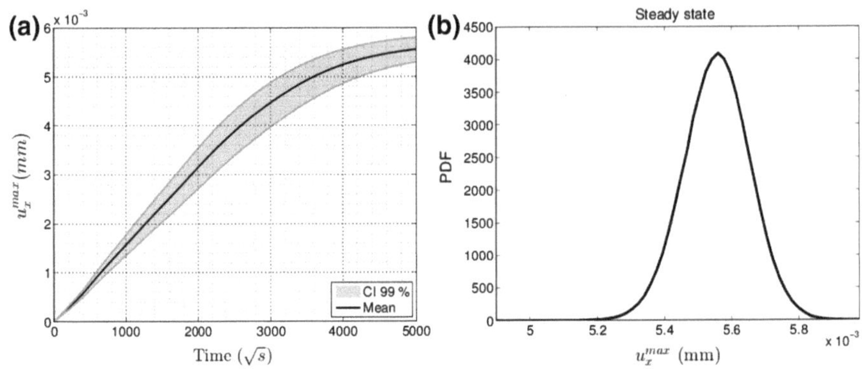

Fig. 17 Evolution of the stochastic process $u_x(\mathbf{x}, t)$ for a particular material point (**a**) PDF of $u_x(\mathbf{x})$ for the same point at the steady state

5 Conclusions

In this chapter, the multiphysics modeling of the hygro-mechanical behavior of heterogeneous composite materials was investigated. We first show the advantage of the Langmuir model which allows a large class of diffusion phenomena to be represented, and can be used when the classical Fick model cannot represent a particular diffusion process. This model is based on a differentiation of the water molecules which can be free or bonded to the polymer network. This model thus allows delay or, on the contrary, fast absorption to be represented, at early instants of diffusion. We showed the efficiency of the model with a numerical study in which an epoxy-based composite reinforced with carbon fibers was analyzed. This analysis was possible through the implementation of the Langmuir diffusion model within a finite element software. In this study, we have shown that a non-Fickian epoxy resin also led to an anomalous diffusion behavior for the composite. The geometrical distribution (regular or random distributions) of the fibers within the material tends to influence the local fields such as the moisture content and the elastic quantities.

The second part of this chapter dealt with the coupling between the diffusion process and the mechanical states. We introduced a dependency of the diffusion coefficient on the local strain field based on the free volume theory. A numerical study was possible owing to the implementation of the nonlinear hygro-elastic problem within finite element software. We studied the impact of this coupling on three-point flexural bending during moisture absorption and compared the results with numerical simulations based on the uncoupled Fick model. We have shown that the external loadings may have a significant impact on the diffusion process. We have also pointed out that such couplings may not modify the shape of the sorption kinetics, which may indeed remain Fickian.

The last part of the chapter was dedicated to uncertainties observed in diffusion kinetics which influence the response of the model. We have proposed to tackle this difficulty using a parametric vision of the uncertainties and modeling the random local and global hygro-elastic response through polynomial chaos expansions. The computation of the stochastic solution was made with an L^2 projection method which only requires the evaluation of the solution for particular realizations of the input random variables. If no analytical solution exists for a given complex hygro-elastic problem, these solutions may be obtained with finite element software. We have highlighted the capabilities and efficiency of the proposed technique with the analysis of a polyamide-based composite reinforced with glass fibers. In this study, both diffusion coefficient and maximum moisture content were random.

References

1. Zhou Z, Benbouzid M, Charpentier JF, Scuiller F, Tang T (2013) A review of energy storage technologies for marine current energy systems. Renew Sustain Energy Rev 18:390–400
2. Yagoubi JE, Lubineau G, Roger F, Verdu J (2012) A fully coupled diffusion-reaction scheme for moisture sorption-desorption in an anhydride-cured epoxy resin. Polymer 53:5582–5595
3. Vaddadi P, Nakamura T, Singh RP (2003) Transient hygrothermal stresses in fiber reinforced composites: a heterogeneous characterization approach. Compos A 34:719–730
4. Shen C, Springer G. Moisture absorption and desorption of composite materials. J Compos Mater 10:2–20
5. Loos AC, Springer G (1979) Moisture absorption of graphite-epoxy composites immersed in liquids and in humid air. J Compos Mater 13:131–147
6. Roy S, Xu W, Park S, Liechti K (2000) Anomalous moisture diffusion in viscoelastic polymers: Modeling and testing. J Appl Mech 67:391–39
7. Crank J (1975) The mathematics of diffusion (2nd edn). Oxford University Press
8. Perreux D, Suri C (1997) A study of the coupling between the phenomena of water absorption and damage in glass/epoxy composite pipes. Compos Sci Technol 57:1403–1413
9. Carter HG, Kibler K (1975) Langmuir-type model for anomalous moisture diffusion in composite resins. Oxford University Press
10. Vaddadi P, Nakamura T, Singh RP (2003) Inverse analysis for transient moisture diffusion through fiber-reinforced composites. Acta Mater 51(1):177–193
11. Bond DA (2005) Moisture diffusion in a fiber-reinforced composite: part I—non-Fickian transport and the effect of fiber spatial distribution. J Compos Mater 39:2113–2129
12. Joliff Y, Belec L, Chailan J (2013) Modifed water diffusion kinetics in an unidirectional glass/fibre composite due to the interphase area: experimental, analytical and numerical approach. Compos Struct 97:296–303
13. Peret T, Clément A, Fréour S, Jacquemin F (2014) Numerical transient hygroelastic analyses of reinforced Fickian and non-Fickian polymers. Compos Struct 116:395–403
14. Fahmy A, Hurt J (1980) Stress dependence of water diffusion in epoxy resin. Polym Compos 1:77–80
15. Yaniv G, Ishai O (1987) Coupling between stresses and moisture diffusion in polymeric adhesives. Polym Eng Sci 27:731–739
16. Marom G, Broutman L (1981) Moisture penetration into composites under external stress. Polym Compos 2:132–136
17. Neumann S, Marom G (1985) Stress dependence of the coefficient of moisture diffusion in composite materials. Polym Compos 6:9–12
18. Neumann S, Marom G (1986) Free-volume dependent moisture diffusion under stress in composite materials. J Mater Sci 21:26–30

19. Youssef G, Freour S, Jacquemin F (2009) Stress-dependent moisture diffusion in composite materials. J Compos Mater 43:1621–1637
20. Caflisch RE (1998) Monte Carlo and quasi-Monte Carlo methods. Acta Numerica 7:1–49
21. Papadrakakis M, Papadopoulos V (1996) Robust and efficient methods for stochastic finite element analysis using Monte Carlo simulation. Comput Methods Appl Mech Eng 134:325–340
22. Ghanem R, Spanos P (1991) Stochastic finite elements: a spectral approach. Springer, Berlin
23. Ghanem R (1999) Ingredients for a general purpose stochastic finite elements implementation. Comput Methods Appl Mech Eng 168:19–34
24. Ghanem R (1999) Stochastic finite elements for heterogeneous media with multiple random non-Gaussian properties. ASCE J Eng Mech 125:24–40
25. Le Maitre OP, Knio OM, Najm HN, Ghanem RG (2004) Uncertainty propagation using Wiener-Haar expansions. J Comput Phys 197(1):28–57
26. Soize C (2010) Identification of high-dimensional polynomial chaos expansions with random coefficients for non-Gaussian tensor-valued random fields using partial and limited experimental data. Comput Methods Appl Mech Eng 199:2150–2164
27. Clément A, Soize C, Yvonnet J (2013) Uncertainty quantification in computational stochastic multiscale analysis of nonlinear elastic materials. Comput Methods Appl Mech Eng 254:61–82
28. Nouy A, Clément A (2010) An extended stochastic finite element method for the numerical simulation of random multi-phased materials. Int J Numer Meth Eng 83(10):1312–1344
29. Abhilash A, Joshi SP, Mukherjee A, Mishnaevsky L Jr (2010) Michromechanics of diffusion-induced damage evolution in reinforced polymers. CompSciTech 71:333–342
30. Neumann S, Marom G (1987) Prediction of moisture diffusion parameter in composite materials under stress. J Compos Mater 6:68–80
31. Wiener N (1938) The homogeneous chaos. Am J Math 60:897–936
32. Xiu DB, Karniadakis GE (2002) The Wiener-Askey polynomial chaos for stochastic differential equations. SIAM J Sci Comput 24(2):619–644
33. Soize C, Ghanem R (2004) Physical systems with random uncertainties: chaos representations with arbitrary probability measure. SIAM J Sci Comput 26(2):395–410
34. Deb M, Babuska I, Oden JT (2001) Solution of stochastic partial differential equations using Galerkin finite element techniques. Comput Methods Appl Mech Eng 190:6359–6372
35. Babuska I, Tempone R, Zouraris GE (2005) Solving elliptic boundary value problems with uncertain coefficients by the finite element method: the stochastic formulation. Comput Methods Appl Mech Eng 194:1251–1294
36. Le Maitre OP, Najm HN, Ghanem RG, Knio OM (2004) Multi-resolution analysis of wiener-type uncertainty propagation schemes. J Comput Phys 197(2):502–531
37. Le Maitre OP, Knio O, Najm H, Ghanem R (2001) A stochastic projection method for fluid flow. i. basic formulation. J Comput Phys 173:481–511
38. Le Maitre OP, Reagan M, Najm H, Ghanem R, Knio O (2002) A stochastic projection method for fluid flow. II. Random process. J Comput Phys 181:9–44
39. Reagan M, Najm H, Ghanem R, Knio O (2003) Uncertainty quantification in reacting flow simulations through non-intrusive spectral projection. Combust Flames 132:545–555

Reliability of Composite Marine Structures

A.J. Sobey, J.I.R. Blake and R.A. Shenoi

Abstract Traditional engineering design takes a deterministic view of the world. This type of analysis has provided a safe method for designing structures for many years. However, reliability analyses have been developed to account for the stochastic loads encountered in service and the variation in geometries and material properties seen after production. There are a range of techniques available for this analysis each with their own advantages and disadvantages. This chapter explores methods for composite analysis using simplified methods and simulation techniques.

Keywords Monte Carlo simulations · Analytical structural analysis · Composite risers · Grillages · Natural fibres · Flax

1 Introduction

The marine industry is making a concerted effort to move from a deterministic to a risk-based approach to design. In an effort to account for the risks it is necessary to determine the probability of an occurrence. Reliability methods have been available for a number of years but are being applied more regularly in industry and much of the literature seeks to provide guidance on how these tools can be used.

Sobey et al. [1] reviewed a number of papers relating to reliability of composite marine structures concluding that many studies focus on simplified structures: plates and cylinders. Recently, more complex marine structures have been analysed by Gaspar [2] and [3], Zayed [4] and [5], Taflandis [6], Silva [7], Luis [8] and Deco [9] but these are focused predominantly on steel structures. Further, to the lack of complexity in the examples being investigated only a small number of limit states are generally being considered, the focus is predominantly on strength and fatigue, normally using a simple S-N approach, and it is proposed that the number of failure

A.J. Sobey (✉) · J.I.R. Blake · R.A. Shenoi
Fluid Structure Interactions Group, University of Southampton, Southampton, UK
e-mail: ajs502@soton.ac.uk

P. Davies and Y.D.S. Rajapakse (eds.), *Durability of Composites in a Marine Environment 2*, Solid Mechanics and Its Applications 245,
https://doi.org/10.1007/978-3-319-65145-3_7

modes being considered need to be increased, with increasing accuracy. This is supported by Sobey et al. [10] who investigated the reliability of marine structures between first principles rules and design rules. This investigation highlighted the importance of the deflection criterion, relative to the strength, for composite marine structures. Since these publications, multi-scale modelling has been increasingly investigated as this provides the opportunity for increased accuracy and the ability to incorporate a wider range of limit states. Sriramula and Chryssanthopoulos [11] quantified the uncertainty in composite materials and reviewed different stochastic methods suggested by researchers. They classified the uncertainties starting at constituent (micro) level to ply (meso) level and component (macro) level. They concluded that multi-scale modelling approaches are promising for stochastic analysis of composite structures. This is because the uncertainties at the micro-level propagate to higher scales and cause significant effects on macro scale stiffness and strength properties. Research proposed by Blanchard et al. [12] on flax fibres highlights a similar problem; while the variation in the strength of the materials is much less variable, there is a much greater variation in properties at the micro-level, making prediction of the more complex limit states difficult.

In the design of marine structures, there are a number of rule sets that take a risk-based approach to design, incorporating some prescribed probability of failures. The International Ship and offshore Structures Congress (ISSC) collated a number of annual probabilities of failure from different codes related to the marine industry, as shown in Table 1. These values represent the annual probability of failure (Pf) for some systems (production and merchant ships) and components (plates, cylinders and shells).

Based on these values and literature, the ISSC has given a list of recommended reliabilities for different types of marine and offshore structures. These values are shown in Table 2.

These values give an idea for the target probability of failures common in marine structures, to allow a comparison to the examples given later. The values given are not as low as might be expected, with probabilities of failure above the level at which society is normally comfortable according to a study performed by Keese and Barton [15] for the aircraft industry.

Table 1 Annual Pf in existing structures [13]

Type of structure	Relevant code	Annual P_f
Production ship	Current Codes	$10^{-4} < 10^{-4}$
Merchant ship	Current Codes	10^{-3}
Cylindrical shells	NPD/DNV, API RP2T	10^{-6}–10^{-4}, 10^{-5}–10^{-4}
Stiffened flat plates	NPD/DNV API RP2T	10^{-5}–10^{-4}
Stiffened panels	API RP2T, RCC/API Bul-2U	10^{-4}
Stiffened plates	API RP2T, RCC/API Bul-2U	10^{-3}
Stiffened shell bays	API RP2T, RCC/API Bul-2U	3×10^{-4}
Offshore structures	API RP2A, LRFD, CSAS471	4×10^{-4}
	CSAS471	10^{-5}–10^{-4}

Table 2 ISSC recommended reliabilities for marine structures [14]

Unit	Failure probability
Monohull	10^{-5}–10^{-3}
Semi-submersibles:	
Hulls	10^{-4}–10^{-3}
Moorings	2×10^{-3}–10^{-2}
Tension leg platforms:	
Hull	10^{-4}–10^{-3}
Tethers	10^{-5}–10^{-4}

This chapter reviews analysis performed at the University of Southampton on grillage top hat stiffened plate structures, typical of boatbuilding practice as an example, using Monte Carlo simulations to analyse different composite materials. The paper analyses the differences in reliability between E-glass/vinylester and carbon/epoxy composite materials using a Navier grillage analysis. This reliability analysis is extended to investigate new materials, in this case flax. Multi-scale analysis is added into the grillage analyses to gain a better prediction of the material properties and the importance of accurate properties is highlighted by assessing the effects of durability issues related to water uptake in composite risers.

2 General Methodology

A number of analyses are performed to demonstrate the benefits of the reliability methodology and to show how different limit states affect the probability of failure. The focus is on simplified analytical structural models in combination with Monte Carlo Simulations. Monte Carlo Simulations provide a more accurate assessment of the probability of failure than FORM/SORM (First Order Reliability Method/Second Order Reliability Method) approximations. However, they are time-consuming requiring in the order of 10^3 more simulations, and therefore more calls of the structural model. These methods become feasible when used in conjunction with rapid analytical solutions with reduced run times. Figure 1 shows the general methodology where the structural analysis method selected and the failure criteria are different depending on the analysis shown below, and are detailed in the relevant sections.

From an assumed statistical distribution and Coefficient of Variation for each basic state variable, a population of values for each basic state variable is randomly generated. Each basic state variable is used in the structural model and so any one combination of randomly generated basic state variables represents a design possibility for structural analysis. A stress and deflection can then be generated and these are compared to the failure criteria, if the criteria are exceeded then a failure occurs. This is then repeated, summing each failure that occurs, until the probability of failure converges; where convergence normally occurs at approximately 10^3 times more simulation runs than the probability of failure.

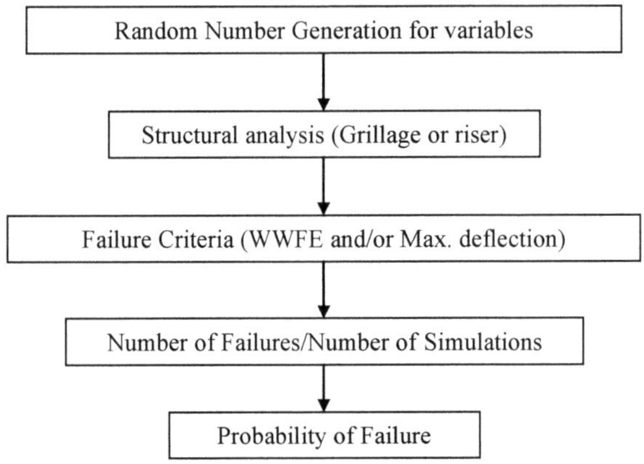

Fig. 1 General Monte Carlo simulation methodology

Table 3 Zinoviev failure criteria [17]

Longitudinal tension failure	$\sigma_1 = X_T$
Longitudinal compressive failure	$\sigma_1 = X_C$
Transverse tensile failure	$\sigma_2 = Y_T$
Transverse compressive failure	$\sigma_2 = X_C$
In-plane shear failure	$\tau_{12} = S_{12}$

Failure criteria are taken from those suggested by the World Wide Failure Exercise (WWFE) [16]. As per these recommendations, the Zinoviev, Tsai and Puck failure criteria have been selected. The Zinoviev failure criteria details are shown in Table 3 which in the case of strength are the ultimate strength values in compression, X_C, or tension, X_T, and in shear, S_{12} (where 1, 2 represent the local in-plane coordinate system with 1 in the longitudinal direction, 2 in the transverse direction).

The Tsai failure criterion [18] is given in Eq. 1, and is a popular criterion where in the cases given the value of F_1 is −0.5 which is the value given as a good approximation for standard composite materials.

$$\left(\frac{\sigma_1}{X_T X_C}\right)^2 + \left(\frac{\sigma_2}{Y_T Y_C}\right)^2 + \left(\frac{1}{X_T} - \frac{1}{X_C}\right)\sigma_1 + \left(\frac{1}{Y_T} - \frac{1}{Y_C}\right)\sigma_2 + \left(\frac{2F_1 2\sigma_1\sigma_2}{\sqrt{X_T X_C Y_T Y_C}}\right) + \left(\frac{\tau_{12}}{S_{12}}\right)^2 = 1, \tag{1}$$

Puck criteria [19] details are given in Table 4 where the strength criteria are similar to the Zinoviev, or maximum strength, criteria with an empirical adaptation in the regions where multiple stresses are applied.

$\epsilon_{1T,C}$ is the failure strain, tensile or compressive, ϵ_1 is the normal strain of a unidirectional layer, ν_{f12} is the Poisson's ratio of the fibres, E_{f1} is the fibre modulus in x1 direction, $m_{\sigma f}$ is the mean stress magnification factor, γ_{21} is the shear strain of a unidirectional layer, σ_{1D} is the stress value for linear degradation, $\rho_{\perp\parallel}^{(-)}$ is the slope

Table 4 Puck failure criteria [19]

Fibre failure in tensión	$\frac{1}{\epsilon_{1T}}\left(\epsilon_1+\frac{\nu_{f12}}{E_{f1}}m_{\sigma f}\sigma_2\right)=1$
Fibre failure in compression	$\frac{1}{\epsilon_{1C}}\left\|\left(\epsilon_1+\frac{\nu_{f12}}{E_{f1}}m_{\sigma f}\sigma_2\right)\right\|=1-(10\gamma_{21})^2$
Inter-fibre failure mode A (for transverse tension)	$\sqrt{\left(\frac{\tau_{12}}{S_{12}}\right)^2+\left(\rho^{(+)}_{\perp\parallel}\frac{Y_T}{S_{21}}\right)^2+\left(\frac{\sigma_2}{Y_T}\right)^2}+\rho^{(+)}_{\perp\parallel}\frac{\sigma_2}{S_{12}}=1-\frac{\sigma_1}{\sigma_{1D}}$
Inter-fibre failure mode B (for moderate transverse compression)	$\frac{1}{S_{21}}\left(\sqrt{\tau_{21}^2+\left(\rho^{(-)}_{\perp\parallel}\sigma_2\right)^2}\right)+\rho^{(-)}_{\perp\parallel}\sigma_2=1-\frac{\sigma_1}{\sigma_{1D}}$
Inter-fibre failure mode C (for large transverse tension)	$\left[\left(\frac{\tau_{21}}{2(1+\rho^{(-)}_{\perp\perp})S_{21}}\right)^2+\left(\frac{\sigma_2}{Y_C}\right)^2\right]\frac{Y_C}{(-\sigma_2)}=1-\frac{\sigma_1}{\sigma_{1D}}$

of the longitudinal fracture envelope and $\rho^{(-)}_{\perp\perp}$ is the slope of the transverse fracture envelope. Other limit states selected for each study are documented in the relevant sections.

3 Grillage Analysis

Composite grillage structures, shown in Fig. 2, are analysed as the exemplar as they predominate in marine structures. The results are demonstrated using analysis taken from Sobey et al. [1]. It utilises a simple grillage theory reported in Vedeler [20].

The deflections for the analysis are generated using Eq. 2,

$$\boldsymbol{w(x,y)}=\sum_{m=1}^{\infty}\sum_{n=1}^{\infty}\boldsymbol{a_{mn}\sin\frac{m\pi x}{L}\sin\frac{n\pi x}{B}},\tag{2}$$

The coefficient a_{mn} is dependent on the flexural rigidities of the stiffeners ($D_{g,\ b}$),

$$\boldsymbol{a_{mn}}=\frac{16\boldsymbol{PLB}}{\pi^6\boldsymbol{mn}\left\{\boldsymbol{m^4(g+1)\frac{D_g}{L^3}+n^4(b+1)\frac{D_b}{B^3}}\right\}}.\tag{3}$$

The moments can then be found in the beams or girders (M_s) from Eq. 5,

$$M_s=-D_s\frac{\partial^2 w}{\partial x^2},\tag{4}$$

and the shear force can also be found for the beams and girders Q_s from Eq. 6,

$$Q_s=\frac{\partial M_s}{\partial x}.\tag{5}$$

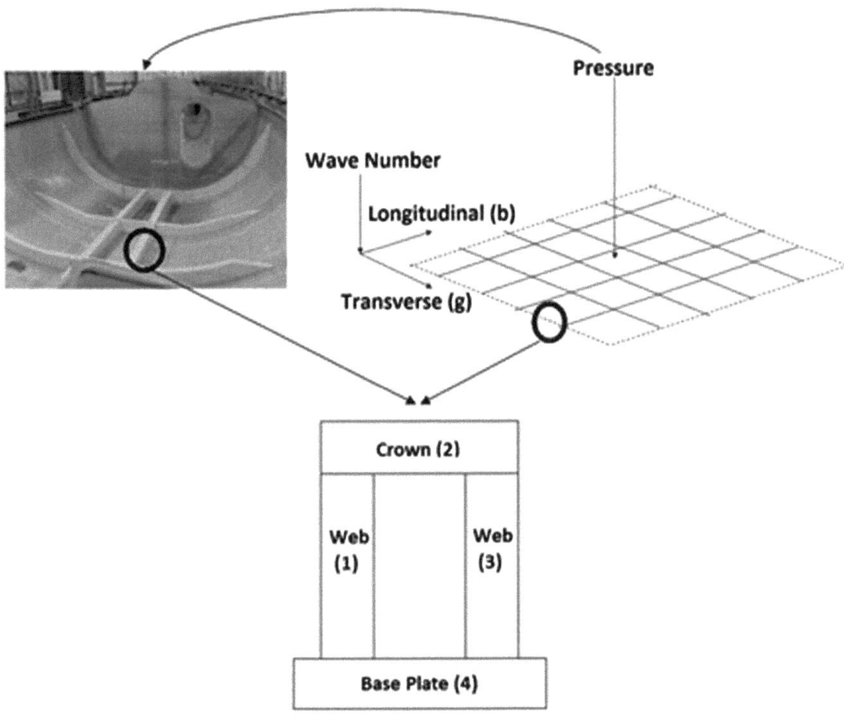

Fig. 2 Grillage topology [21]

Table 5 Verification of grillage method [1]

Property	Established	Current
Clarkson—Deflection (mm)	9.63	9.87
Clarkson—Stress (MPa)	165.52	170.13
Datoo—Shear at Crown Corner (MPa)	99.00	98.72
Datoo—Shear at Neutral Axis (MPa)	101.00	102.76

This grillage theory is verified as shown in Table 5. The grillage stresses and deflections are compared to orthotropic plate method as simulated by Clarkson [22] and the shear forces are compared to Datoo [23]. The Clarkson panel has a length and width of 3810 mm. The panel consisted of four beams and girders with dimensions 254 mm deep 127 mm wide with 18.288 mm thick flanges and 9.144 mm thick webs (the unusual significant figures are an artefact of imperial to SI unit conversion). A pressure of 137.9 kPa was applied to each panel and this panel is used in further studies, unless stated. The results show a close match to the previously reported results.

Table 6 Demonstration of accuracy of grillage method [24]

Method			FEM	Grillage method	Difference (%)
Deflection (mm)			26.23	20.75	−20.9
Stress (MPa)	Top layer	σ_x	225.9	255.4	13.1
		σ_y	2.051	2.31	12.6
		σ_{xy}	0.43	0	–
	Bottom layer	σ_x	262.5	290.9	10.8
		σ_y	2.33	2.63	12.9
		σ_{xy}	0.40	0	–

Table 7 Comparison of the reliability between glass and flax fibre reinforced structures [1]

	Carbon/epoxy	E-glass/vinylester
Probability of failure	6×10^{-4}	3×10^{-2}
Normalised compared to mass	1.07×10^{-6}	4.19×10^{-5}
Normalised compared to cost	1.77×10^{-7}	1.1×10^{-5}

Yang et al. [24] compared the answers from the grillage theory with the addition of classical laminate plate theory, giving a good comparison to Finite Element Method, also allowing the incorporation of layer by layer stresses, shown in Table 6.

This analysis shows that the grillage gives an accurate approximation to the finite element analysis, while benefitting from low computational expense. This allows an accurate interpretation of the reliability without the need for surrogate modelling. It will also allow for extension to co-varying input values where two variables are linked and changes in one will elicit changes in the other.

4 Standard Composites

Material selection can be a difficult process in a number of industries. More expensive materials can give improved performances with the ability to develop structures with lower mass but determining the additional benefits can be a difficult task. In Sobey et al. [1] a simple grillage analysis is used, without the addition of classical laminate plate theory, showing a comparison of reliability of carbon/epoxy and E-glass/vinylester. All of the WWFE criteria are assessed for first failure as well as a grillage deflection limit state, taken as 10% of the stiffener spacing. The results of this analysis are shown in Table 7.

The results show that the carbon provides a lower probability of failure than the E-glass composite, which is the expected result. The analysis shows the importance of the deflection criterion; all of the failures that do occur are against the deflection criteria; in neither case are there any strength failures. This provides difficulty in performing reliability analysis for composite materials as the analysis is application specific. Therefore, the deflection criterion must be clearly defined and is often found from design rules or "rule of thumb". This makes comparisons of results difficult as the criterion changes with each application.

5 Extension to Multi-scale Reliability

Current continuum methods allow a good estimation of the structural integrity. However, the types of modelling performed in the previous section have limitations. To allow for improvements in estimating the behaviour of composites, understanding the fibre, the matrix and their interaction at the smaller scale is important to improve accuracy and understand damage initiation. A grillage can be assessed at different levels for example, as shown in Fig. 3, providing an ability to assess a wider range of limit states and more accurate material properties. This can be achieved by building a macro level model from representative building blocks of fibre/matrix elements. These building blocks or unit cell approaches depend upon a representative volume element (RVE) to define the micro-level interaction. A comparison between the reliability of a grillage analysis using material properties developed from the Halpin-Tsai method for the composite and that from the more accurate unit cell or Representative Volume Element (RVE) approach, Mutlu et al. [25], is shown below.

First, a unit cell model or RVE is developed for graphite/epoxy, Fig. 4, where the material properties are shown in Table 8.

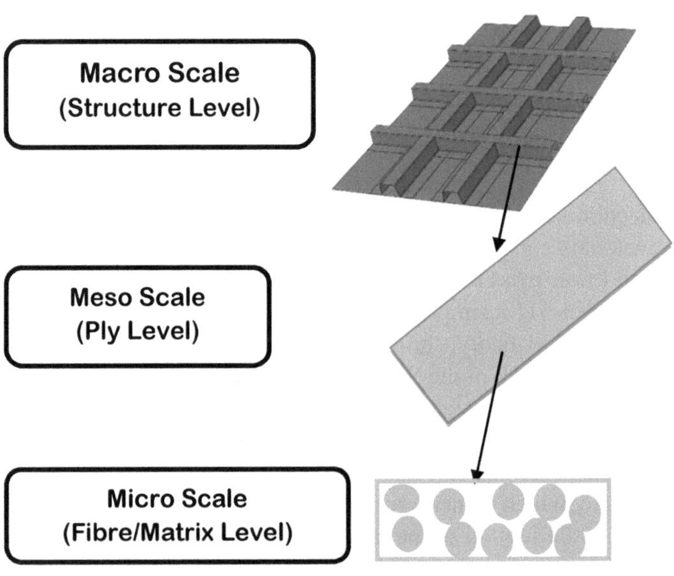

Fig. 3 Methodology for multi-scale modelling [25]

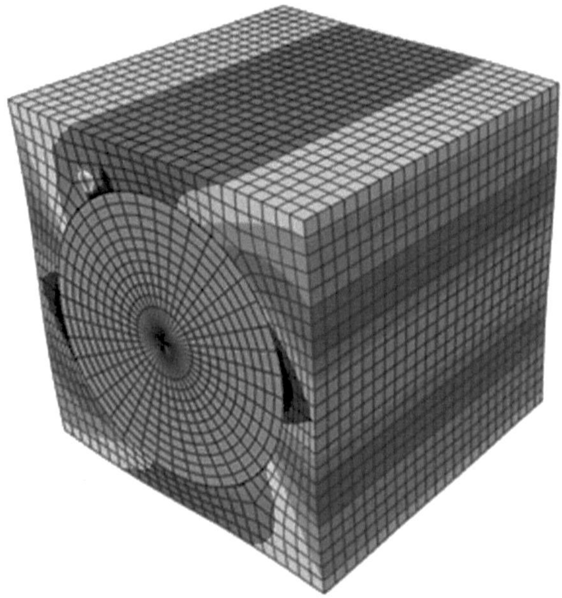

Fig. 4 RVE model under periodic boundary conditions

Table 8 Material properties for AS4 and 3501-6

	E_1 (GPa)	E_2 (GPa)	G_{12} (GPa)	ν_{12}	ν_{23}
Graphite AS4	235	14	28	0.2	0.25
Epoxy 3501-6	4.8	4.8	1.8	0.34	0.34

The RVE, shown in Fig. 4, is created with the diameter of the fibre calculated to ensure a fibre volume fraction of 0.6. Periodic boundary conditions are often chosen for approximating large systems and are used at the edges to represent the connection with the rest of the layer. The periodic boundary condition means that deformations on opposite surfaces are the same and that the stress vectors acting on opposite sides are opposite in direction.

The results are compared to Sun [26] and the experimental results from Daniel and Lee [27] and Sun and Zhou [28] shown in Table 9. The results show a good correlation for most of the properties. The Young's modulus shows similar results to previous FEA and experiments in both directions. The shear modulus, however, gives a close approximation to the previous finite element analysis, but sits between the two sets of experiments in the G_{12} direction and underestimates one set of experimental results in the G_{23} direction. The Poisson's ratio is also similar to the previous FEA modelling, underestimating the experiments. The latter sets of results are judged to be similar enough and differences may be caused by experimental error, as well as some inaccuracy between the modelling and experiments.

Table 9 Verification of material properties against graphite/epoxy composite

Elastic constants (GPa)	RVE model [25]	FE model [26]	Experiment [27, 28]
E_1	143.2	142.6	142/139
E_2	10	9.6	10.3/9.85
G_{12}	5.84	6.0	7.6/5.25
G_{23}	2.93	3.1	3.8
v_{12}	0.24	0.25	0.3
v_{23}	0.31	0.35	–

The RVE is approximated using a surrogate due to the high computational expense; it is not possible to run it for each step in the Monte Carlo Simulation. The surrogate modelling includes four main steps:

- Initiation of data points using a sampling plan;
- perform FEA at each sample point;
- construct the surrogate from the sample points;
- explore the surrogate to ensure accuracy.

In order to define the sample points, Latin Hypercube is used, for which Forrester et al. [29] describes an efficient method. To construct the surrogate model, Kriging is used based on Eq. (6)

$$\varphi^{(i)} = \exp\left(-\sum_{j=1}^{k} \theta_j \left|x_j^{(i)} - x_j\right|^{p_j}\right), \tag{6}$$

where $\varphi^{(i)}$ are the parameters used to fit the model, k is the number of design variables, x_j are the components of the sample points; 40 sample points are used to construct the surrogate and another 20 test points are used to check the accuracy of the model. Average prediction errors are used to check the accuracy of the surrogate model: the calculation of the longitudinal Young's modulus is used as an illustrative example in Fig. 5 to show the accuracy of the predictions. Young's modulus of the surrogate is generally slightly over predicted compared to that of the complete FE analysis, with an average error of 3%.

The averages predicted for the errors are presented in Table 10 with the other material properties giving lower predicted errors.

The verification studies of the RVE-surrogate model show that it can be used with confidence to implement the material properties into the reliability analysis. A composite stiffened plate with 0.55 fibre volume fraction is analysed as an example of typical marine structures. For this analysis glass fibre is used with an epoxy resin with material properties given in Table 11, the coefficient of variation for the material properties are also included and each is assumed to follow a Normal distribution.

Four variables are employed to calculate the layer properties within the surrogate model: E_f, E_m, G_f and G_m. The stiffened plate is defined in Table 12, which also

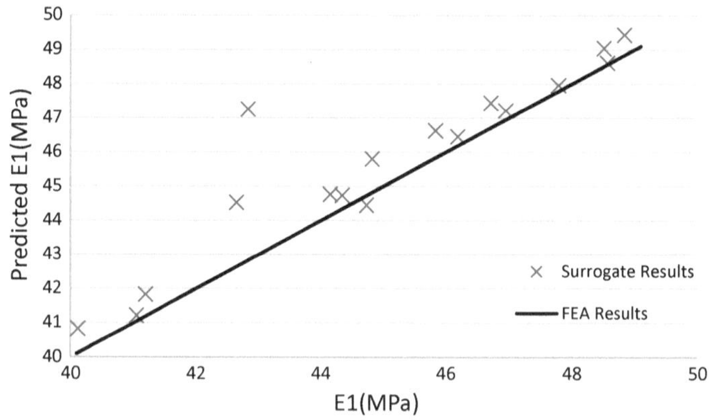

Fig. 5 Surrogate model control for E1

Table 10 Accuracy of surrogate model [25]

	Avg. Pred. error (%)
E_1	3.05
E_2	0.12
G_{12}	0.003

Table 11 Micro-level variables and coefficients of variation (CoV)

	E (GPa)	v	CoV (%)
Glass	73.1	0.22	3
Epoxy	3.45	0.35	3

Table 12 Macro level variables

Variable	Definition	Value (mm)	CoV (%)
L	Length	3072	3
B	Breadth	3072	3
P	Pressure (kPa)	110	3
a	Crown width	211	–
b	Crown thickness	12.92	–
c	Web-Flange thick	12.92	–
d	Web height	258	–
e	Flange width	232	–
	Ply thickness	1.615	–
	Number of plies	8	–

shows the coefficient of variation for the variables all of which are varied using a normal distribution.

Two approaches are applied in this study: first, the material properties and effects of the constituent properties are calculated by the micro-level RVE modelling; second, the material properties are calculated using the analytical approximation of

Table 13 Probability of failure results [25]

	RVE-Surrogate	Halpin-Tsai
Equivalent material properties	0.0855	0.146
Layer by layer material properties	0.0856	0.146

Halpin-Tsai from the constituent properties. For both approaches, the material properties are assigned to be equivalent for each layer, with one material property for the laminate, and this is compared to an analysis where each ply is varied using the same random process as defined in the Monte Carlo simulation but instead of assigning one material property for the entire thickness each layer has a different set of properties. The probability of failure is assessed against selected criteria: the maximum deflection criterion (L/150) is taken from Lloyds Register Rules for Special Service Craft [30] and the Zinoviev failure criterion is selected from those recommended by the World Wide Failure Exercise. The results are shown in Table 13 where 10^8 runs were performed for each case. It is observed that there is not a significant difference with the assignment of material properties at each layer giving similar probabilities of failure for both cases.

However, a significant difference is observed for the probability of failure calculated between the RVE-Surrogate and Halpin-Tsai approaches. One of the reasons for this could be differences in the material properties estimated by the surrogate model. The longitudinal Young's modulus estimated by the surrogate model is higher than the RVE model calculations. The number of sample points could be enhanced to have better estimations.

This study discusses multi-scale approach for the reliability analysis of composite stiffened plates. Micro and macro level modelling is applied separately to be able to include micro- and macro variables into the reliability analyses. Representative volume element approach and Halpin-Tsai analytical approximation are performed in order to evaluate the material properties. In general, good agreement is achieved between the RVE model and the literature. The material properties obtained from a square packed RVE seem adequate in predicting the material properties of composites. Surrogate modelling using Kriging also gives good estimates for the material properties. There are some small differences in the predictions of the longitudinal Young moduli for the test cases but it is used as a surrogate model of the finite element RVE model with good accuracy and low computational cost. Since this study is ongoing work, the accuracy of the RVE and surrogate model can be improved. At macro level model, the grillage method provided the required accuracy and speed to be used in a probabilistic analysis. As a future work, the RVE model will be studied in detail by including the microstructural damages such as voids, defects and delamination to investigate the macro level responses. However, a complete set of material properties is required, including distributions and coefficients of variation.

6 Flax Composites

The RVE approach is successful in providing an estimate of material properties, and could be extended to understand a wider range of limit states. Reliability analysis can still provide an ideal way to include the variation of materials into the structural level analysis, and explore the feasibility of new materials. Blanchard et al. [12] use a grillage analysis to explore the reliability of natural composites. Flax fibres are well investigated at the fibre scale, Blanchard [30], but there are limited studies to show its potential as a structural material. A number of studies indicate that the lack of uptake is due to the variability of the materials at the fibre scale, but Blanchard et al. show that this is not the case at the laminate scale. Reliability analysis is therefore used to compare the flax to E-glass materials, the standard composite material most likely to be replaced by flax, to understand how flax composites perform. A grillage is outlined in Table 14 with two stiffeners in each direction. The structure used is similar to the previous grillage (Sect. 5) but with a lower load and a smaller grillage. Due to the disparity in mass and thickness for the flax materials and the E-glass, two studies are performed: one to simulate a mass-constrained problem common to most engineering situations, and second a volume-constrained condition which is commonly found in the boatbuilding industry.

Material properties, and their variations, are collated from previous experiments and literature and are shown in Table 15. It can be seen that whilst the variations for the materials at the laminate level are comparable to each other, the stiffness and strength of the flax are much lower. Low stiffness for flax plies is counteracted at the laminate level as flax is about 20% thicker than the E-glass laminates for the same number of plies. Flax laminates also have a much lower mass, meaning more plies can be used before mass equivalence with the glass laminates.

Results for the mass- and volume-constrained cases are shown in Table 16 subjected to a strength limit state (Zinoviev). The probability of failure for the E-glass composite grillage is zero (i.e. no predicted strength failures). A grillage constructed from flax/epoxy to the same volume as that of the E-glass/vinylester has a relatively large probability of failure, 0.146, showing that the flax is not a feasible replacement. For the mass-constrained applications the flax provided a safe option, judged to have a reasonable probability of failure compared to the ISSC criteria.

Table 14 Grillage topology for E-glass and flax [12]

Definition	E-glass/Vinylester	Flax/epoxy (same volume)	Flax/epoxy (same mass)
Crown width (mm)	211	211	211
Crown thickness (mm)	12.92	12.93	20.06
Web thickness (mm)	12.92	12.93	20.06
Web height (mm)	258	258	258
Flange width (mm)	211	211	211
Flange thickness (mm)	12.92	12.93	20.06

Table 15 Material properties for flax and glass fibre reinforced composites

Materials	Flax/epoxy			E-glass/vinylester		
Mechanical properties	Mean	CoV (%)	Ref.	Mean	CoV (%)	Ref.
Longitudinal Young's modulus (MPa)	8 180	5.08	31	29 200	6.16	33
Transverse Young's modulus (MPa)	8 180	5.08	31	23 900	7.95	33
Poisson's ratio	0.25	5.00	N/A	0.16	6.25	33
Shear modulus (GPa)	2.07	5.31	32	4.5	6.67	33
Tensile strength in the fibre direction (MPa)	90.9	7.89	31	512.5	4.39	33
Compressive strength in the fibre direction (MPa)	90.32	4.76	32	363.4	20.64	33
Tensile strength perpendicular to the fibres (MPa)	90.9	7.89	31	350.9	2.54	33
Compressive strength perpendicular to the fibres (MPa)	90.32	4.76	32	336.4	7.49	33
Shear strength (MPa)	38.01	5.81	32	44.7	1.34	33
Ply thickness (mm)	0.539 mm wet for 221 g/m^2		31	0.417 mm wet for 610 g/m^2		33
Composite density (kg/m^3)	1220		31	1980		33

Table 16 Comparison of the reliability between glass and flax fibre reinforced structures [12]

	E-glass/vinylester	Flax/epoxy (same volume)	Flax/epoxy (same mass)
Probability of failure (P_f)	0	0.146	3×10^{-6}
Normalised compared to mass (P_f/kg)	0	7.9×10^{-4}	1.05×10^{-8}
Volume of the grillage (m^3)	0.145	0.145	0.234
Mass of the grillage (kg)	286.96	176.96	285.00
Mass of the fibres (kg)	216.55	78.32	125.4
Mass of the resin (kg)	70.41	99.13	159.7
Thickness of the laminate (mm)	12.92	12.93	20.05

In this case the thickness of the flax laminate is much higher, 20 mm compared to 13 mm, but the number of layers remains similar to the E-glass laminate with an increase from 31 to 37. This highlights that whilst flax might not provide a like for like structural material it might be feasible in applications where sustainability is an important driver. A key element to this analysis is that the deflection is not used in the limit state; this is especially relevant as deflection values for the flax are high. Incorporating the maximum deflection criterion (L/150) used in the previous study, which is taken from Lloyds Register Rules for Special Service Craft, would have led to much higher probability of failures and required much higher masses and volumes.

The study indicates that flax fibre composites might be beneficial for lightly loaded structural applications with mass driven constraints and a willingness to integrate sustainable materials. At the yarn scale, the variability in mechanical properties is high but is comparable to conventional composites at the laminate level. A higher fibre volume fraction will significantly improve flax properties. Some key material properties are still missing for natural fibres and a complete understanding of the structural potential of natural composites will require a more complete set of material properties and limit states. The multi-scale modelling approach for reliability predictions will become more powerful for these uncertain materials, as they will be better defined at the unit cell level.

7 Durability of Composite Risers

Composite risers are built to operate in harsh conditions for long periods of time. During this period the material properties degrade and it is at this point that failures are most likely to occur. The previous studies, as with much of the literature, focuses on material properties after production and undamaged. This study aims to investigate the difference in probability of failure for structures with moisture damage. This is done using a static catenary model outlined in his section.

The governing mathematical equations for the catenary shape requires the following inputs: w, weight per unit length of the cable; T_H, horizontal tension; ρ_l, density of constituent laminate structure; ρ_i, density of internal fluids; ρ_f, density of external medium; h, operational depth; and the geometry of the cylindrical pipe to evaluate of the second order differential equation, Eq. 7,

$$y''(x) = \frac{1}{\alpha}\sqrt{1 + [y'(x)]^2}; \tag{7}$$

where α is the catenary parameter, $\frac{T_H}{w}$. The boundary conditions are imposed such that

$$y(0) = 0, \tag{8}$$

$$y'(0) = 0, \tag{9}$$

$$\tan\theta_0 = 0; \tag{10}$$

where θ_0 is the angle of the seabed at the touchdown point. Assuming the angle of the seabed is 0 leads to Eq. 11,

$$y(x) = \frac{T_H}{w}\left[\cosh\left(\frac{w}{T_H}x\right) - 1\right] - h. \tag{11}$$

which is a catenary curve with a departure point at $y = 0$, sea level; and a touch-down point at a depth of h metres below the sea surface.

The length of the catenary section, s, based on the horizontal, x, direction can be determined using Eq. 12,

$$s(x) = \frac{T_H}{w}\sin\text{h}\left(\frac{w}{T_H}x\right)\cdot \tag{12}$$

Forces acting on the riser include tension, both vertical and horizontal, as well as the bending moments and were estimated using Eqs. 13 and 14,

$$T_v(x) = w.s(x), \tag{13}$$

$$T(x) = \sqrt{\left(T_v^2(x) + T_H^2\right)}, \tag{14}$$

where $T_v\,(x)$ is the distribution of the vertical component of tension along the length of the riser. In addition, the bending moments acting on a segment of the riser can be estimated by multiplying the curvature by the bending stiffness inherent to the material and geometry as shown in Eq. 15,

$$M(x) = EI.\kappa(x) = EI\frac{w}{T_H\cosh^2\left(\frac{w}{T_H}x\right)}. \tag{15}$$

The local analysis was conducted to determine the stresses and strains acting on the laminate structure of the riser. This was done using Classical Laminate Theory (CLT), using the notation from Nijhof [34], in Eq. 16,

$$\begin{pmatrix} N \\ M \end{pmatrix} = \begin{bmatrix} A & B \\ B & D \end{bmatrix}\begin{Bmatrix} \varepsilon^0 \\ \kappa \end{Bmatrix} \tag{16}$$

where A is the extension stiffness matrix; B is the bending-extension coupling effects between in-plane stresses and curvatures and between bending and twisting moments and in-plane strains; D is the stiffness of the laminate in the perpendicular direction under the influence of bending and twisting moments; ε^0 is the vector of strain at a particular point in the laminate; κ is the vector of curvatures induced in the laminate by the external forces; N represents the in-plane forces acting on the segment of riser and M representing the corresponding induced moments.

The tensions experienced by the riser can be expressed, in Eq. 17, as the resultant forces (N_x, N_y, N_{xy}) where

$$N = (T_V, 0, 0), \tag{17}$$

with $N_{xy} = 0$. Likewise, the bending moments occur with respect to the y-direction, in Eq. 18, thus

$$M = (M_B, 0, 0). \tag{18}$$

The in-plane strain vector $\{\varepsilon\}_{k}$- for the kth lamina is given by Eq. 14,

$$\{\varepsilon\}_k = \{\varepsilon^0\} + z_k\{\kappa\}. \tag{19}$$

The stress-strain relationship can be determined for the kth lamina by employing Eq. 20,

$$\{\sigma\}_k = [Q']_k\{\varepsilon^0\} + z_k[Q']\{\kappa\}, \tag{20}$$

where Q' is the transformed reduced stiffness matrix for each lamina, dependent on the lamina's angle relative to the principal direction of the laminate, and $\varepsilon = \{ \varepsilon_x, \varepsilon_y, \varepsilon_{xy})$ is a vector of in-plane strains experienced by the laminate. The stress components of $\{\sigma\}_k$ are evaluated by the Tsai-Wu failure criterion and used to determine the reliability.

The Tsai-Wu criterion is chosen as a good predictor of first-ply failure. The static catenary equation is compared to a static analysis performed by Zhan [35] using Riflex™ (Riser System Analysis Programme) with beam elements, the results for tension are shown in Fig. 6.

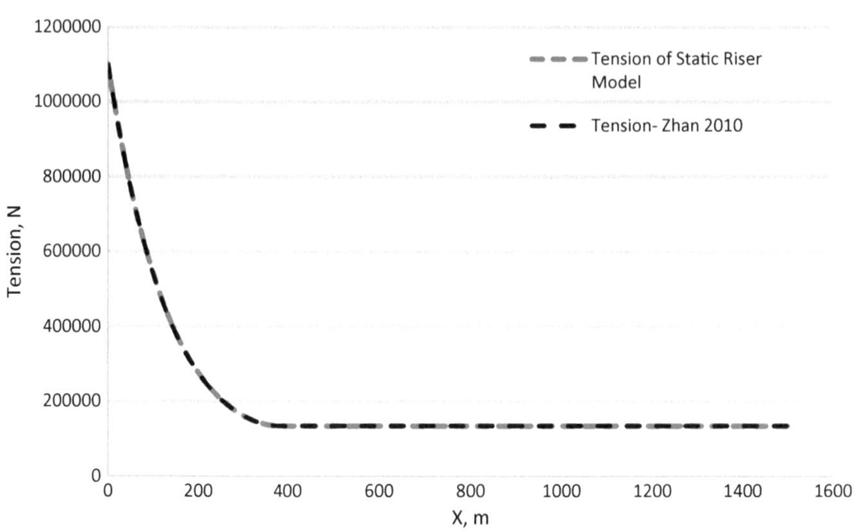

Fig. 6 Comparison of tension forces from static global model to Zhan [35]

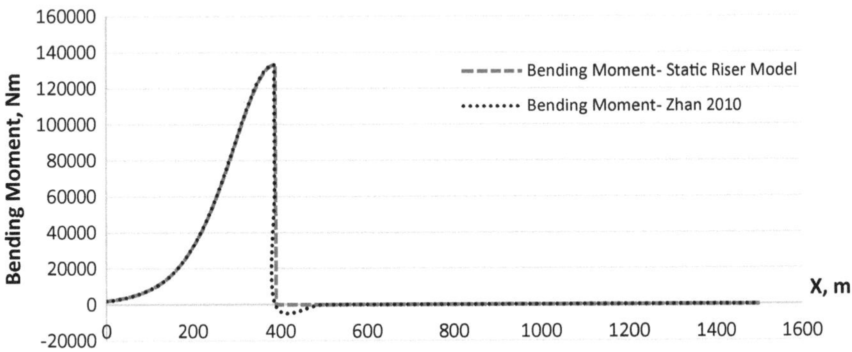

Fig. 7 Comparison of bending moments from static global model to Zhan [35]

Table 17 Topology for the riser

Number of laminae	56-ply
Composite material	Carbon/Epoxy
Layup	$[90/15/-15/90/45/-45/45/-45/45/-45]_5 + [45/-45]_3$
Thickness of steel liner (m)	0.007
OD (m)	0.22
Ply thickness (m)	0.000075
Laminate wall thickness (m)	0.0112
Weight/metre (N)	1030

The results show that there is a good match between the two results. A comparison of the bending moments is shown in Fig. 7.

The bending moments provide a similar accuracy to the tensions, but near the touchdown point there are some discrepancies. A reliability analysis is performed on a riser with properties given in Table 17, based on those given in [36].

The variables for the analysis are given in Table 18 with material properties taken from [36] and the coefficients of variation are taken from [37].

The reliability analysis is performed for three sets of material properties. The first is the intact material properties assuming no damage from moisture uptake. The second is the properties given by Malmstein et al. [38] on glass/epoxy composites in freshwater. The properties of these materials gave a flexural strength reduced by 50% and the weight increased by 2%. The third set of properties come from experiments developed by carbon/epoxy specimens by Zafar et al. [39] who found an increase in weight of 2.12%. The corresponding loss of tensile strength was 20% and Young's modulus was reduced by 10%. These values replace the properties given in Tables 17 and 18. It is considered that Zafar et al. [38] represents the more realistic values for risers due to the similar materials and tensile properties alongside experiments conducted in salt water. Due to the lack of available data, Malmstein et al. [38] also used to represent a more pessimistic case as the results are so

Table 18 Riser predicted material variations

	Composite	
	Mean [35]	CoV [36] (%)
Xt (MPa)	2450	5
Xc (MPa)	1570	16
Yt (MPa)	70	18
Yc (MPa)	133	16
S (MPa)	98	11
Ex (GPa)	125	10
Ey (GPa)	9.588	16
G (GPa)	5.4	20
ν	0.35	11

Table 19 Comparison of riser reliability with and without material degradation

Depth (m)	Composite		
	Normal	Malmstein [38]	Zafer [39]
1500	2.00E-06	2.80E-05	2.00E-06
2000	1.00E-06	1.84E-04	5.00E-06
3000	1.30E-05	3.12E-03	6.60E-05
4000	1.25E-04	3.02E-02	5.33E-04

dissimilar to each other, as they are taken in freshwater which gives a higher water uptake than salt water (Table 19).

These results show that determining the effects of durability are potentially important for these applications. At higher depths there is a larger separation in results between the intact and moisture reduced properties. Whilst analysis at current depths shows a small separation it is important that, as greater depths are explored, these properties are accounted for. The different degraded properties change the probability of failure with the more realistic Zafer properties decreasing the probability of failure at greater depths, but making less of a difference at lower depths of operation. There can be a difference in the moisture uptake depending on a number of factors, including which surfaces are wetted and the heat of the material. Risers have a large temperature differential and are likely to only be exposed to moisture on one side. It is therefore important to be able to more accurately determine the durability properties of these materials, and specifically their applications.

8 Conclusions

This chapter reviews work using reliability analysis for composites structures using Monte Carlo simulations. Results from a rapid analytical grillage method and a riser are compared, showing the benefits in using simplified methods to gain a good understanding of the structural problem. The grillage analysis highlighted how reliability analysis can be used to help make design decisions. It also showed the importance of the deflection criteria for composite applications. The growing importance of multi-scale modelling is highlighted and it is shown how differences in prediction can be made, using these models to provide additional accuracy and eventually more limit states to the analysis. Also reviewed is a comparison of glass fibre composite grillage structures with a natural fibre reinforced composite structures with two approaches: "same volume" and "same mass". The results show that the probability of failure of natural composites is lower than for conventional fibre structures while also requiring a structure with a larger volume. However, due to the low density of natural fibre reinforced composites, it can be applicable to secondary or tertiary structure with light mass requirements. Finally the effects of material durability are incorporated in the models and their importance is shown. At low depths, the difference in probability of failure is small, but as the operating depth increases the importance of these properties also increases. There is a large difference between the more realistic and pessimistic material properties and further work needs to be performed to judge the degradation of properties in these applications. Key themes throughout all of the analyses are the material properties, their distributions and the values by which they vary. To develop good distributions takes a large quantity of experiments, which often are not performed or treated with the statistical care that would make these sets more useful; this is especially the case when looking at the co-variance. Furthermore, the properties that are developed are for intact materials, not those that have been in operation for a number of years, and the latter are the more realistic properties for when failure is likely to occur. Virtual experiments, using micro-scale modelling to degrade the properties, are likely to provide an opportunity to more easily explore these types of properties and an area of importance for the next steps in reliability modelling.

References

1. Sobey AJ, Blake JIR, Shenoi RA (2013) Monte Carlo reliability analysis of tophat stiffened composite plate structures under out of plane loading. Reliab Eng Syst Saf 110:41–49
2. Gaspar B, Naess A, Leira BJ, Guedes Soares C (2002) System reliability analysis of stiffened panel under combined uniaxial compression and lateral pressure loads. Struct Saf 39:30–43
3. Gaspar B, Guedes Soares C (2013) Hull girder reliability using a Monte Carlo based simulation method. Probab Eng Mech 31(1):66–75
4. Zayed A, Garbatov Y, Guedes Soares C (2013) Time variant reliability assessment of ship structures with fast integration techniques. Probab Eng Mech 32(1):93–102
5. Zayed A, Garbatov Y, Guedes Soares C (2013) Reliability of ship hulls subjected to corrosion and maintenance. Struct Saf 43(1):1–11
6. Taflanidis AA, Vetter C, Loukogeorgaki E (2013) Impact of modeling and excitation uncertainties on operational and structural reliability of tension leg platforms. Appl Ocean Res 43(1):131–147
7. Silva JE, Garbatov Y, Guedes Soares C (2014) Reliability assessment of a steel plate subjected to distributed and localized corrosion wastage. Eng Struct 59(1):13–20
8. Luís RM, Teixeira AP, Guedes Soares C (2009) Longitudinal strength reliability of a tanker hull accidentally grounded. Struct Saf 31(1):224–233

9. Decò A, Frangopol DM, Zhu B (2012) Reliability and redundancy assessment of ships under different operational conditions. Eng Struct 42(1):457–471
10. Sobey AJ, Blake JIR, Shenoi RA (2014) A rapid method for reliability analysis of composite tophat stiffened structures using a first principles method and design rules. Trans R Inst Nav Archit—Int J Marit Eng 156:1–28
11. Sriramula S, Chryssanthopoulos MK (2009) Quantification of uncertainty modelling in stochastic analysis of FRP composites. Compos A Appl Sci Manuf 40(11):1673–1684
12. Blanchard JMFA, Sobey AJ, Blake JIR (2016) Assessing the feasibility of natural composite for structural applications. In: 17th European Conference on Composite Materials, Munich, Germany
13. Moan T, Berge S (1997) Design Principles and Criteria, International Ship and Offshore Structures Congress Committee IV.1, Trondheim, Norway
14. Moan T, Berge S (1997) Structural Design of Floating Production Systems, International Ship and Offshore Structures Congress Committee V.6, Trondheim, Norway
15. Keese DL, Barton WR (1982) Risk assessment and its application to flight safety analysis. In Sandia National Laboratories, SAND89
16. Soden PD, Kaddour AS, Hinton MJ (2004) Recommendations for designers and researchers resulting from the world-wide failure exercise. Compos Sci Technol 64:589–604
17. Zinoviev PA, Lebedeva OV, Tairova LP (2002) A coupled analysis of experimental and theoretical results on the deformation and failure of composite laminates under a state of plane stress. Compos Sci Technol 62:1711–1723
18. Kuraishi A, Tsai SW, Liu KKS (2002) A progressive quadratic failure criterion, Part b. Compos Sci Technol 62:1683–1695
19. Puck A, Schurnmann H (2002) Failure analysis of frp laminates by means of physically based phenomenological models. Compos Sci Technol 62:1633–1662
20. Vedeler G (1945) Grillage beams in ships and similar structures. Grondahl & Son
21. Sobey AJ, Blake JIR, Shenoi RA (2013) Implications of failure criteria choices on the rapid concept design of composite grillage structures using multiobjective optimisation. Struct Multidiscip Optim 47:735–747
22. Clarkson J (1965) The elastic analysis of flat grillages. Cambridge University Press
23. Datoo MH (1991) Mechanics of fibrous composites. Elsevier Applied Science, London
24. Yang N, Das PK, Yao XL (2010) Reliability analysis of stiffened composite panel. In: Asranet Conference, Glasgow
25. Mutlu U, Sobey AJ, Blake JIR, (2016) Multi scale reliability approach to analyse composite marine structures. In: 2nd International Conference on Safety and Reliability of Ships, Offshore and Subsea Structures SAROSS 2016, Glasgow, UK
26. Sun CVR (1996) Prediction of composite properties from a representative volume element. Compos Sci Technol 57:171–179
27. Daniel IM, Lee JW (1990) Progressive transverse cracking of crossplycomposite laminates. J Compos Mater 24:1225–1243
28. Sun CT, Zhou SG (1988) Failure of quasi-isotropic composite laminates with free edges. J Reinf Plast Compos 7:515–557
29. Forrester A, Sobester A, Keane A (2008) Engineering design via surrogate modelling. John Wiley & Sons Ltd, West Sussex
30. Lloyd's Register (2015) Lloyd's Register Rules and Regulations-Rules and Regulations for the Classification of Special Service Craft, Rulefinder Ver 9.23 (Jan 2015)
31. Blanchard JMFA, Sobey AJ, Blake JIR (2016) Multi-scale investigation into the mechanical behaviour of flax in yarn, cloth and laminate form. Compos B 84:228–235
32. Yan L (2012) Effect of alkali treatment on vibration characteristics and mechanical properties of natural fabric reinforced composites. J Reinf Plast Compos 31(13):887–896
33. Tekalur SA, Shivakumar K, Shukla A (2008) Mechanical behavior and damage evolution in E-glass vinyl ester and carbon composites subjected to static and blast loads. Compos B Eng 39:57–65

34. Nijhof AHJ (1993) Analysis of laminated composites. In: Shenoi RA, Wellicome JF (eds) Composite materials in maritime structures: Fundamental Aspects, vol 1, Cambridge University Press, Cambridge
35. Zhan JP (2010) Review and verification of marine riser analysis programs : global response analysis. Norwegian University of Science and Technology thesis
36. Pham DC, Su Z, Sridhar N, Qian X, Huang Z, Sobey AJ, Shenoi, RA (2016) Experimental and numerical studies of large-scaled filament wound T700/X4201 composite risers under bending. In: 17th European Conference on Composite Materials, 2016, Munich, Germany
37. Philippidis TP, Lekou DJ, Aggelis DG (1999) Mechanical property distribution of CFRP filament wound composites. Compos Struct 45(1):41–50
38. Malmstein M, Chambers AR, Blake JIR (2013) Hygrothermal ageing of plant oil based marine composites. Compos Struct 101:138–143
39. Zafar A, Bertocco F, Schjodt-Thomsen F, Rauhe JC (2012) Investigation of the long term effects of moisture on carbon fibre and epoxy matrix composites. Compos Sci Technol 72:656–666

Multiscale Modelling of Environmental Degradation—First Steps

Andreas T. Echtermeyer, Abedin Gagani, Andrejs Krauklis and Tobiasz Mazan

Abstract Fibre-reinforced plastics are praised for their good corrosion resistance. However, when this resistance needs to be documented for safety-critical applications extensive and time-consuming test programmes are needed. A better quantitative understanding of the fundamental mechanisms behind degradation should help to reduce the testing effort. A multiscale approach for modelling degradation in water and hydrocarbons is described. The environmental degradation happens on the scale of the constituent materials: fibres, matrix and sizing (interface). Local concentration profiles of the fluid inside the material need to be known to predict degradation. The global engineering properties are then calculated from the constituents using finite element analysis and homogenisation. Describing degradation with the multiscale approach is a promising method for reducing the current test effort. Much more work is needed to create enough confidence in the models that then can be used for designing real components. The first steps are described here showing the models and how they can be connected.

Keywords Composites · Environmental degradation · Multiscale modelling · Failure mechanisms · Marine · Offshore

1 Introduction

Composites (fibre-reinforced plastics) are frequently recommended for having good corrosion resistance. They do not corrode like metals and that is certainly an advantage. Corrosion of metals has been studied for many years, it is well understood for common exposures, but corrosion of metals is also still one of the main challenges when using metals in the marine and offshore environment [1–3]. Composite materials can potentially replace metals and solve the corrosion problems. However, composites can degrade with time. The degradation is typically a

A.T. Echtermeyer (✉) · A. Gagani · A. Krauklis · T. Mazan
Norwegian University of Science and Technology, NTNU, Trondheim, Norway
e-mail: Andreas.Echtermeyer@ntnu.no

P. Davies and Y.D.S. Rajapakse (eds.), *Durability of Composites in a Marine Environment 2*, Solid Mechanics and Its Applications 245,
https://doi.org/10.1007/978-3-319-65145-3_8

different process than corrosion, but it still has to be accounted for. If designed and used properly, composites can be operated without maintenance for long times, their entire design life. Service experience is good. But it has been time-consuming and costly to demonstrate for safety-critical applications, that the material properties of composites do not degrade significantly [4, 5].

Today's methods for obtaining long-term properties are based on accelerated testing [6]. There are basically two ways to accelerate testing: increasing the test temperature over the service temperature or increasing the loading above service loads. Increasing the temperature is often not possible, because the composites in the oil and gas industry are already being used close to the maximum temperature that the matrix can tolerate. For marine applications (ships) using Arrhenius-based acceleration to simulate long-term performance may be an option, because the service temperature in water or in the air is significantly lower than the maximum temperature the composite can be tested at. The API 15S standard [7] also uses Arrhenius-type acceleration for its axial testing of end fittings on thermoplastic pipes.

The most widely used method for determining degradation by accelerated testing is long-term testing based on increasing the load beyond the operational loads while exposing the samples to the extreme environments that could be experienced during service. Specimens are tested in saturated conditions and the liquid needs to diffuse into the samples before testing. For large test specimens, this conditioning process can take several weeks, adding to the already long testing times. A 4-mm thick glass fibre epoxy sample can take 8–9 months to saturate with water at 60 °C. Full-scale specimens can typically not be tested in the service environments, because the conditioning process to reach saturation would take many years.

The resistance to short-term high loads is simply measured by short-term static tests to rupture of the conditioned specimens. Comparing the strength of the conditioned specimens with the strength of dry unconditioned specimens shows the reduction of the strength. If the stresses due to extreme operational loads times the required safety factors are still lower than the strength of the conditioned specimens, the design is acceptable in that respect. This approach is only valid if fairly instant degradation mechanisms or just swelling during the conditioning period cause the strength reduction. If chemical degradation would slowly lead to further damage with time, it would not be detected. Typically, fairly inert constituent materials are chosen to prevent the effect of long-term degradation due to chemical attack.

The long-term properties of the conditioned specimens themselves are typically measured by cyclic fatigue testing or stress rupture tests, to demonstrate resistance against cyclic or long-term constant loads, respectively. Test results at short times and high loads are then extrapolated to long times and low loads. The low loads would be the operational loads times safety factors. Testing of the preconditioned specimens has to be done in the conditioning environment to ensure that the fluids do not diffuse out of the specimen during the long-lasting test. If the test periods are very short and diffusion out of the specimen is slow the testing can possibly be done just in the air. However, it should be noted that diffusion out of the specimens could happen quite rapidly, it often occurs at a comparable rate with that of diffusion into the specimens.

For SN curves test data are typically obtained between 20 and 1,000,000 cycles. An example is given schematically in Fig. 1. The test points are fitted by linear regression giving the mean SN curve. A characteristic curve is established by shifting the mean curve to lower numbers of cycles. Data can also be extrapolated to longer cycle times. The characteristic curve is then shifted down further to account for the uncertainty in the extrapolation process. The length of extrapolation is also limited. The approach for obtaining stress rupture curves is basically the same, except that numbers of cycles are replaced by time. Testing times for a design life of 25 years are up to 10000 h (a bit more than a year). The approach described here is in accordance with DNV-OS-C501 [9] and DNVGL-RP-F119 [8]. Very similar approaches are given in API 17 J [10], API 15S [7] and ASTM D 2992 [11]. This method has worked well in many applications.

An inherent assumption in the extrapolation process is that no change in failure mechanisms or phase change in the material happens during the extrapolated time, i.e. the maximum measured time to the end of design life. Based on the good experience with the methods this assumption seems to be right for the composite materials and environments used today. But it is good to keep in mind that these are statistical methods that do not say anything about the physics and processes involved during fatigue or stress rupture. For example, early polyethylene pipes showed a phase transformation in stress rupture after a little more than 10000 h. The stress rupture curve got steeper, not allowing direct extrapolation of stress rupture data measured only below 10000 h [12].

The lack of understanding of the physics of long-term degradation is reflected in the large test programmes required. Figure 2 shows schematically SN curves for different testing conditions. The reference curve is usually the dry material at room temperature. If the temperature gets increased or the material gets soaked with water or hydrocarbons, the SN curve usually drops. Since we do not know how much it

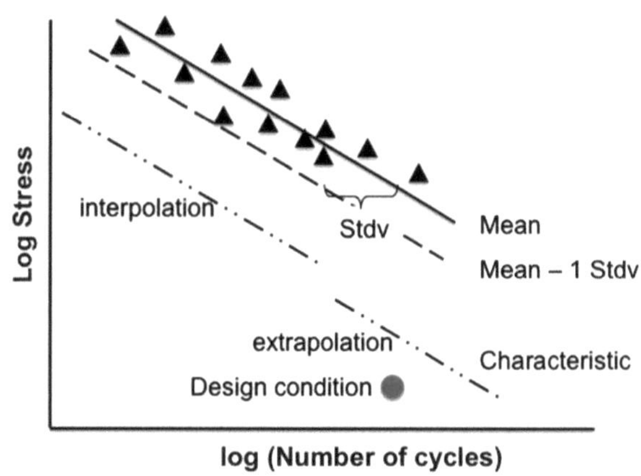

Fig. 1 Typical SN *curve* and extrapolation of data to higher number of cycles. Based on a figure in [8]

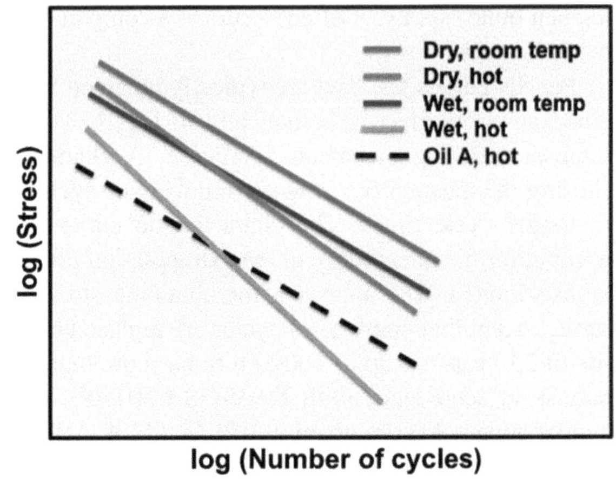

Fig. 2 Expected changes of an SN curve due to different environments

drops, this has to be measured. The drop in static strength compared to dry conditions could be used as a shift factor. But we also do not know if the slope of the curve changes, requiring measurement of a full new SN curve. Every new environmental condition requires obtaining a new SN curve. If we would understand the mechanisms better and could quantify them (even in a conservative way), testing could be significantly reduced.

Trying to understanding long-term degradation on the ply or laminate level has so far not led to procedures that can reduce the test effort. Much qualitative reasoning has been developed, but still new testing is required for new materials and new environments. This is partially due to the complexity of composites. Fibres can be arranged in many different ways, they can be straight, woven, knitted. They may be even distributed or in bundles with resin rich areas between them. The laminate may have many voids, thermal stresses, not be fully cured, etc. It seems that a model addressing the main effects and not ignoring the details that can be applied to most composite laminates has not been found yet.

This chapter proposes addressing degradation with the multiscale approach. In particular, environmental degradation should happen on the level of the constituent materials: fibres, matrix and sizing. The sizing on the fibres is governing the strength between fibres and matrix and needs to be included as a constituent material, even though commonly only fibres and matrix are mentioned as constituent materials. In the next paragraphs we will explain how degradation can affect the constituent materials. Based on the properties of the constituent materials it should be possible to predict ply properties and, further, laminate properties. Predicting laminate properties from ply properties is well established through laminate theory. Elastic properties and static strength can be well predicted. A more detailed analysis would show that strength and fatigue behaviour may be influenced by stacking sequences and other laminate specific properties, but we will ignore these effects in the "first steps" described here. More challenging is the prediction of ply properties from constituent properties. Elastic properties can be quite well predicted by "classical micromechanics" [e.g. 13, 14, 15], but strength is not as predictable. More recent work using Finite Element (FE) simulations of Representative Volume Elements (RVE) are more promising in also predicting strength on the ply level [e.g. 16, 17].

2 Degradation of Reinforcing Fibres

We limit ourselves here to degradation in hydrocarbons and water. Carbon fibres are typically considered to be inert to both liquids. They would not absorb or react with either liquid. Carbon fibres also do not change their properties in the temperature range of interest for marine and offshore applications ranging from about −60 °C up to 200 °C. Exposure to other harsh chemicals, such as H_2S may affect carbon fibres. In such cases, new measurements of the properties of the fibres would have to be made.

Unprotected glass fibres exposed to water lose their strength quickly, especially if they are also mechanically loaded [18, 19]. This is due to the glass material partly dissociating into ions and due to ion exchange in the aqueous medium, weakening the glass. If the matrix protects the glass, this effect is largely reduced. A rule of thumb is that strength and fatigue properties of E-glass fibres drop by 10% if protected by the matrix. But it depends on the composition of the glass fibre and some recent glass fibres perform better than traditional E-glass fibres [19]. For safety-critical applications the rule of thumb is not sufficient and measurements are still required. It is also not clear whether the slopes of SN curves or stress rupture curves change due to exposure to water. Simple hydrocarbons are seen not to affect the properties of the glass fibres.

For a micromechanics model, carbon fibres can be assumed to have the same properties in air at room temperature as in any typical marine or offshore environment. This is a convenient situation for developing models. For glass fibres, the degradation of material surrounded by the matrix needs to be understood. It should depend on the water content that reaches the fibre, the stress of the fibre and the temperature. Research is planned by the authors to investigate this aspect in more detail.

Organic fibres, such as aramid, can show more complicated behaviour. They can absorb liquids and their internal microstructure can be changed. These fibres are not covered in this document.

3 Degradation of the Matrix

Degradation of the matrix can be described by chemistry and polymer science. We use polyamide 11 (PA11, Nylon 11) as an example of how degradation can be described by a multiscale approach. A detailed description was given by Mazan

[20–22]. PA11 can react with water via hydrolysis. The reaction equation is given by Jacques' kinetic model [21, 23]:

$$r(W) = \frac{dn}{dt} k_H[Amide][Water] - k_R[Acid][Amine]$$

The reaction rate r is given by the amide bonds being broken with time *dn/dt*. This depends on concentrations of amide groups, water and the hydrolysis products —amine and acid groups. Note that the reaction constants k_H and k_R have a temperature dependence given by an Arrhenius law. This means that the concentration and temperature profile of water in the component have to be known.

Based on the reaction equation, the change in molecular weight with time can be calculated. Figure 3 shows a comparison between calculation and experimental results for water exposure at 120 °C. An equilibrium is reached at about 70 days.

The change in properties is given here by basic chemical reaction kinetics. It is also interesting to see that hydrolysis creates equilibrium after 70 days. The equilibrium is governed by the balance between hydrolytic degradation of amide bonds and repolymerization of its products.

Applying various aspects of polymer science, it was shown that the reduction of molecular weight happens in the amorphous phase. The low molecular weight components migrate towards the crystalline regions of the semi-crystalline PA11 and become part of the crystalline phase. This means crystallinity increases due to degradation of the amorphous phase. It increased from 22 to 35% in the equilibrium condition for this example.

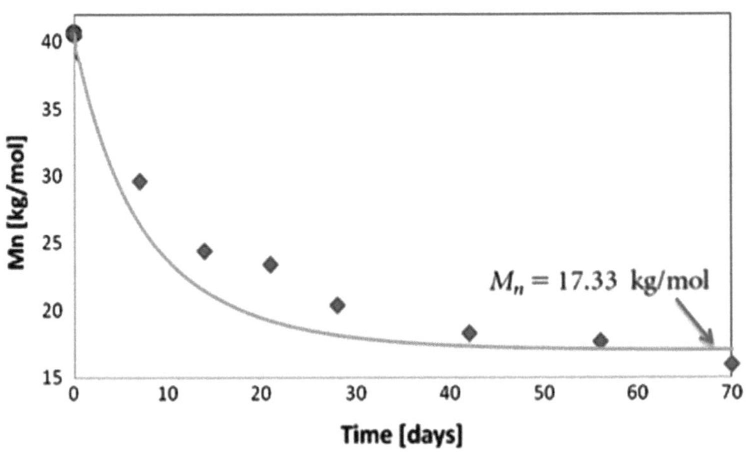

Fig. 3 Change of Molecular weight of PA11 with time at 120 °C (*solid line*). Experimental values are shown as symbols, based on [20]

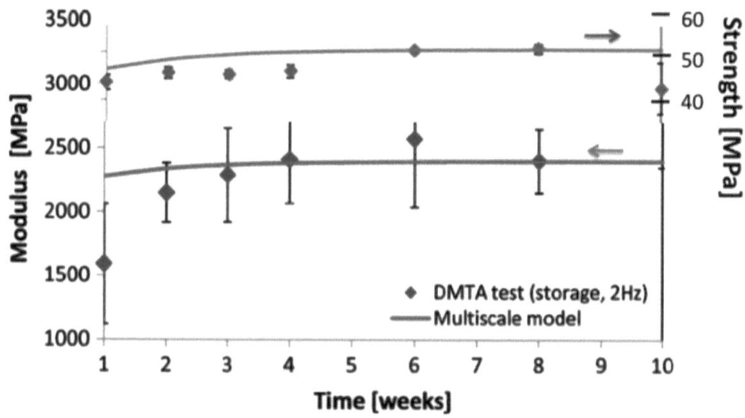

Fig. 4 Change of modulus and strength of PA11 with time at 120 °C. Experimental values are shown as symbols, based on [22]. The modulus was measured by DMTA, the strength was measured using dogbone-shaped coupons

Calculating various polymer properties, it was eventually possible to predict the change of the Young's modulus, as shown in Fig. 4. Making some assumptions that modulus is linked to strength, the strength of PA11 could be predicted as well, as also shown in Fig. 4.

The modelling approach shown here has given fairly good agreement with experiments. Many assumptions were made and it will probably take some time until enough confidence is established to use such methods in design calculations and standards. However, it shows that degradation can be predicted based on first principles of polymer science combined with some empirical molecular correlations. It seems to be a promising way to describe degradation.

Degradation of other matrix materials or in other environments can, in principle, be described in a similar way.

- An equation needs to be found that describes the chemical reactions.
- The influence of the chemical reaction on the molecular structure needs to be established.
- The molecular structure needs to be linked to elastic properties and strength properties.

It is also important to point out that in many matrix materials used in composites in the marine and offshore industry, chemical reactions are not expected to happen. In these cases, strength properties are simply influenced by swelling.

Not all steps are always known. New experimental techniques may have to be developed to establish all relevant correlations. But this approach can bring the prediction of degradation away from statistical curve fitting to a more fundamental understanding.

4 Degradation of Sizing

Degradation of the sizing can be described in the same way as degradation of the matrix. Unfortunately, the detailed chemistry of the sizing is not easily available and, in any case, more complicated molecules are involved. This makes establishing the steps described above more difficult. More empirical methods may be needed to establish the properties of the sizing. They could be fibre matrix pull-out tests [24–26], micro-debonding tests [27–29], etc. These tests are complicated to perform, but they could give very useful input to the degradation.

5 Obtaining Concentration Profiles

The degradation of the matrix and, to a so far unknown extent, of the fibres is dependent on the concentration of the water or hydrocarbons. It is therefore important to integrate into a degradation model the ability to know the concentration profile of the medium.

Many empirical results are available and also a few analytical methods for predicting fluid diffusion in composites. Most use equations proposed by Shen and Springer [30], based on the Fickian law [31]. However, as pointed out by Woo and Piggott [32], the relations fail to precisely predict the effect of fibre volume fraction and fibre orientation on diffusion. In particular, for the diffusion along the fibre direction they predict a lower diffusivity than the one observed experimentally. The reason is that diffusion along the fibres and through the sizing happens more rapidly than through the matrix alone [32]. This is supported by some experimental studies. Using Atomic Force Measurements AFM and Micro Thermal Analysis μTA, Mallarino et al. [33] and then Joliff et al. [34] measured the mechanical and thermal properties of the interphase, the few μm thick layer around each fibre. They reported lower stiffness and softening temperature, attributed to a locally less dense crosslinking and to some plasticization. They consequently deduced that diffusivity is also different locally, higher than in the bulk matrix. In order to obtain a precise diffusion prediction, Joliff et al. also proposed a 2D FE diffusion modelling approach taking the effect of the interphase into account. They modelled the interphase as a few μm thick area around each fibre and obtained good agreement with experimental results [35]. It was important to use a realistic fibre architecture to get more precise results.

The authors extended this idea to a 3-D FE model. They prepared some samples with fibres in the in-plane direction and some samples with the fibres in the out-of-plane directions. This model was able to predict the higher diffusivity in the samples with fibres in the out-of-plane direction, which was not possible to explain

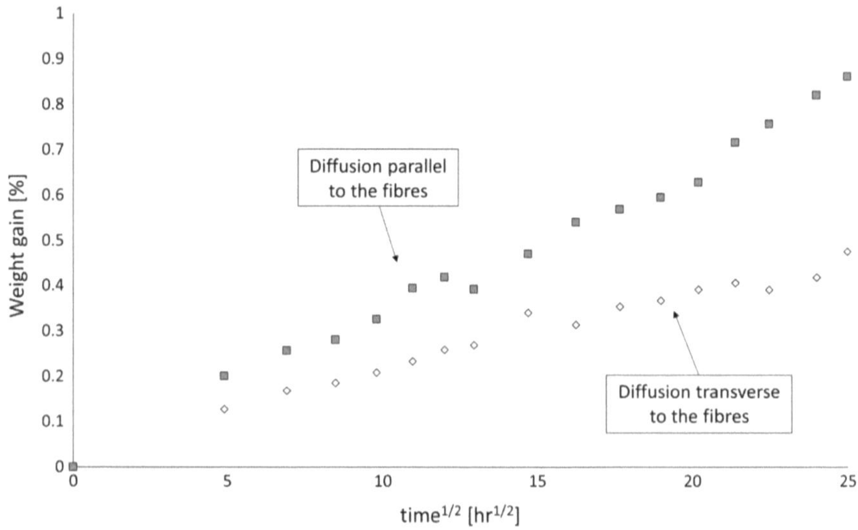

Fig. 5 Anisotropic diffusion of water in a glass epoxy laminate

with the previous models proposed by Shen and Springer [30]. Furthermore, it was also possible to explain the effect of volume fraction on fluid diffusion by means of the same model. Figure 5 shows some preliminary measurements of water uptake in a glass fibre epoxy laminate along the fibre direction and across the fibre direction. Modelling results have also been obtained [36].

In order to highlight the importance of this effect, a problem that composite riser designers face is reported here: a composite riser is immersed in water and terminated with metal end fittings at the ends. The strength of the connection between end fitting and riser pipe is difficult to predict due to stress concentrations, unknown interface properties and dominance of through the thickness stresses. Typically, the design is based on saturated degraded material properties, but this approach may be too conservative. Parts of the laminate inside the end fitting may stay dry during the entire design life due to the long diffusion paths towards the inside of the end fitting. Using this approach will only be possible if good confidence in the diffusion calculations is obtained.

It should be noted that the assumption of the validity of Fickian diffusion has been questioned in some cases. Unexpected increases in diffusion after long exposure times were found, where the water content would have been expected to be constant due to saturation [37, 38]. Voids, applied stresses and matrix cracks are also expected to influence the diffusion characteristics of the laminate. These aspects need to be studied more to gain confidence in the prediction of concentration profiles of fluids in composite components.

6 Linking Constituent Properties to Ply Properties

Once the properties of the degraded constituent materials are known the properties of the composite need to be predicted. As described in the introduction, there are challenges in predicting composite laminate properties from constituent properties, because all the possible interactions are difficult to describe in simple models. This is a challenge that composite science has been facing for many years. For example, the offshore standard DNV-OS-C501 [9] and the Recommended Practice DNVGL-RP-F119 [8] require both that material properties are mainly established on the ply level. Laminate and component properties can subsequently be calculated and a few confirmation tests are required to confirm that the more complicated components and structures are properly described. This approach is widely used and typically referred to as the testing pyramid. Figure 6 shows the testing pyramid from DNV-RP-F119. This RP has a clause that allows also obtaining material properties on the constituent level, provided it can be shown that ply properties are properly described. This is an extension of the testing pyramid to a smaller scale while following the basic philosophy of confirmation testing on the higher level. No guidance is given in the RP on how the link from constituent properties to ply properties shall be done.

A promising approach is to model the microstructure of the ply as a Representative Volume Element RVE. Figure 7 shows such an element as an example. In order to obtain a good estimation of the effective properties, it is necessary to apply boundary conditions that guarantee a compatible deformation field. The RVE is described by a FE mesh and the elastic (or plastic) and strength

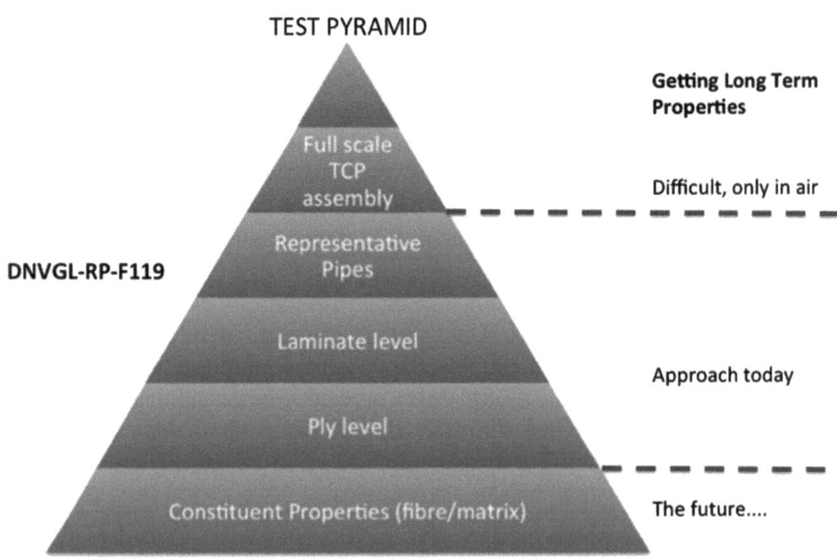

Fig. 6 Test pyramid, as described in DNVGL-RP-F119 [8]

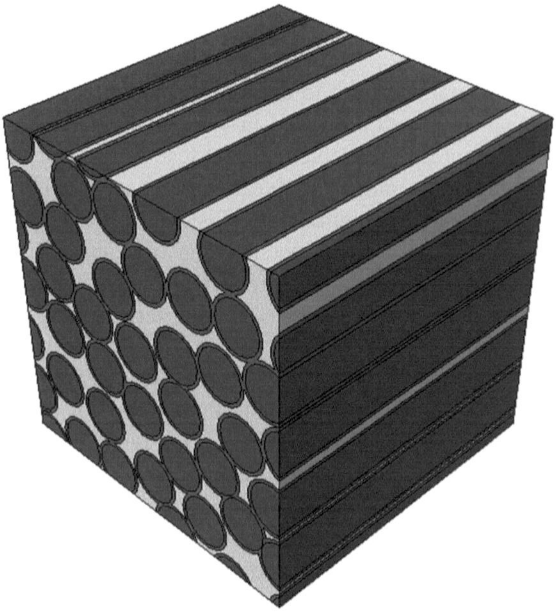

Fig. 7 Possible representative volume elements to describe composite's behaviour from the properties of the constituents

properties of fibres, matrix and sizing (interface) can be put into the respective elements of the mesh. Deformations and failure can then be predicted from the behaviour of the RVE. Failure can be defined in various ways. It can be the maximum stress in a simulated stress strain curve or it can be the occurrence of a certain type and size of damage in the RVE. Since RVEs are not suitable for modelling large structures the properties of the RVE need to be homogenized. Homogenized properties can then be used as ply properties in standard FE codes. Good examples can be found in [39–42].

We are here mainly interested in using the modelling to predict degradation. Unfortunately, all uncertainties in going from constituent properties to homogenized ply properties become part of this modelling approach. For modelling degradation, it may be sufficient to predict the relative change of properties using measurements on the constituent level and scale actual properties on the ply level accordingly. This simplification would require experimental verification, but it could circumvent some of the challenges involved in predicting properties using RVEs and homogenisation. It should be easier to predict relative changes than absolute changes.

The sections above describe how static properties could be obtained from the degraded constituent properties. There is no generally accepted method to relate stress rupture and fatigue to constituent properties. Some work is ongoing by the authors to address this issue, but it is too preliminary to report this now.

7 Discussion

The general approach to describe degradation of composites is based on long-term testing. It is well described in various standards and widely used in industry. The main commercial problem is that testing times, extent and cost are too high. Test programmes running for a year or more are often not feasible in the tight schedules of development programmes. This can lead to composites not being selected for an application, traditional metals being used instead. From a scientific point of view the large test programmes reflect the lack of knowledge behind the mechanisms causing degradation. Extensive long-term testing is compensating for this lack of knowledge.

Understanding the failure mechanisms behind degradation could solve the lack of knowledge and reduce the testing effort. It is probably the only approach that can be taken to achieve the goal of less testing and also shorter testing times. The multiscale approach described here provides a means to characterize degradation on the fundamental level of constituent properties and to link the results to commonly used global engineering properties. The goal to understand composite behaviour based on fundamental properties and principles is as old as composites science. Due to this all the steps described here have been studied to some extent before, and this document has mentioned only a small fraction of the work that has been done. The multiscale approach has, however, been more developed to understand composite behaviour in general. The link to degradation is not established as much. The title of this paper states "first steps" and it is hoped that we will fairly quickly reach a better understanding of degradation in this context.

From an industrial perspective, it will be important that the approach is general, applicable to many materials and reliable. The results will have to be included in design standards, otherwise the old way of extensive long-term testing will remain. Reliability of the results is paramount for this. Predictions of the new models will have to be extensively tested against the old approach until reliability can be assured. The methods employed by the testing pyramid approach of confirming results on a higher level can be applied here. A certain generality and simplicity would also help for the implementation. Finding the major mechanisms that govern the behaviour of degradation and describing them in a widely applicable way will be more important than to characterize all effects involved, provided the simplifications are conservative. Over time, the models can be improved giving more optimal results and confidence.

8 Conclusions

The current way of obtaining long-term degradation results is based on large and time-consuming test programmes. The methods are based on accelerated testing, mainly by increasing loads and having specimens conditioned with the relevant environments. Extrapolation to the longer service times is made by statistical mathematics without a direct relation to the physics involved. The methods have worked well giving good service records and are described in various marine and offshore standards. It would be desirable to shorten the test programmes.

A possible way to shorten test times is a multiscale-based approach. Degradation is described by chemical reaction kinetics, which is well linked to the physical process taking place on a molecular level.

Examples are shown of how engineering properties for the constituent properties can be obtained from the molecular degradation events. Many assumptions are involved in such calculations, but with time confidence could be gained in the methods.

A further step is to obtain ply properties, as they are the typical input to FE design programmes used by engineers. Using Representative Volume Elements to model the interactions of the constituents and subsequent homogenisation of properties seems to be the best approach to obtain the engineering properties.

Since all degradation is dependent on the concentration of the fluids inside the components a good understanding of the diffusion process is also needed. Anisotropic characteristics of the diffusion process should be considered.

The principles of applying the multiscale approach to degradation have been shown. Details and especially experimental evidence that the approach is reliable are still missing. However, it seems to be the most promising approach to reduce testing programmes for documenting degradation in the future.

Acknowledgments This work is part of the DNV GL led Joint Industry Project "Affordable Composites" with nine industrial partners and the Norwegian University of Science and Technology (NTNU). The authors would like to express their thanks for the financial support by The Research Council of Norway (Project 245606/E30 in the Petromaks 2 programme).

References

1. Revie RW (2008) Corrosion and corrosion control. Wiley
2. Revie RW, Uhlig HH (2011) Uhlig's corrosion handbook, vol 51. Wiley
3. Schumacher MM (1979) Seawater corrosion handbook, corrosion of metals in marine environments—an overview. Noyes Publications, Berkshire
4. Salama MM, Stjern G, Storhaug T, Spencer B, Echtermeyer A (2002) The first offshore field installation for a composite riser joint, OTC 14018, offshore technology conference, Houston, Texas, 6–9 May 2002
5. Storhaug T, Echtermeyer AT, Sund OE, Salama MM (2002) Composite tethers—qualified for ultra deep waters, deepwater offshore technology (DOT). Bjørn Paulshus, New Orleans
6. Echtermeyer AT (2014) Integrating durability in marine composite certification. In: Durability of composites in a marine environment. Springer, Netherlands, pp 179–194
7. API 15S (2006) Qualification of spoolable reinforced plastic line pipe, 1st edn, Mar 2006
8. DNVGL-RP-F119 (2015) Recommended practice for thermoplastic composite pipes, Rev.0
9. DNV-OS-C501 (2013) DNV offshore standard "Composite Components" Rev.03
10. API 17 J/ISO 13628-2 (2008) Specification for unbonded flexible pipe, 3rd edn, July 2008

11. ASTM D2992-12 (2012), Standard practice for obtaining hydrostatic or pressure design basis for "Fiberglass" (Glass-Fiber-Reinforced Thermosetting-Resin) pipe and fittings
12. Kinloch AJ, Young RJ (1983) Fracture behaviour of polymers. Gaube E, Kausch HH (1973) Bruchtheorien bei der Industriellen Anwendung von Thermoplasten und glasfaserverstarkten Kunststoffen. Kunststoffe German Plastics, 63(6):391–397
13. Halpin JC (1984) Primer on composite materials: analysis. Technomic Publishing Co, Lancaster, PA
14. Halpin JC (1984) Primer on composite materials: analysis. Technomic Publishing Co, Lancaster
15. Tsai SW (1988) Composite design, 4th edn, Think Composites
16. Naya F, González C, Lopes CS, Van der Veen S, Pons F (2017) Computational micromechanics of the transverse and shear behavior of unidirectional fiber reinforced polymers including environmental effects. Compos Part A: Appl Sci Manuf 92:146–157
17. Saeb S, Staenman P, Javili A (2016) Aspects of computational homogenization at finite deformations: a unifying review from reuss' to voigt's bound. Appl Mech Rev
18. Schutte CL (1994) Environmental durability of glass-fiber composites. Mater Sci Eng R: Rep 13:265–323
19. Renaud CM, Greenwood ME (2005) Effect of glass fibres and environments on long-term durability of GFRP composites. In: Proceedings of 9 EFUC meeting, Wroclaw, Poland
20. Mazan T, Berggren R, Jørgensen JK, Echtermeyer A (2015) Aging of polyamide 11. Part 1: evaluating degradation by thermal, mechanical and viscometric analysis. J Appl Polym Sci 132(20)
21. Mazan T, Jørgensen JK, Echtermeyer A (2015) Aging of polyamide 11. Part 2: General multiscale model of the hydrolytic degradation applied to predict the morphology evolution. J Appl Polym Sci 132(41)
22. Mazan T, Jørgensen JK, Echtermeyer A (2015) Aging of polyamide 11. Part 3: multiscale model predicting the mechanical properties after hydrolytic degradation. J Appl Polym Sci 132
23. Jacques B, Werth M, Merdas I, Thominette F, Verdu J (2002) Hydrolytic ageing of polyamide 11. 1. Hydrolysis kinetics in water. Polymer 43:6439–6447
24. Pitkethly MJ, Favre JP, Gaur U, Jakubowski J, Mudrich SF, Caldwell DL, Drzal LT, Nardin M, Wagner HD, Di Landro L, Hampe A, Armistead JP, Desaeger M, Verpoest I (1993) A round-robin programme on interfacial test methods. Compos Sci Technol 48:205–214
25. Hampe A, Kalinka G, Meretz S, Schultz E (1995) An advanced equipment for single- fibre pull-out test designed to monitor the fracture process. Composites 26:40–46
26. Jones TP, Thorvaldsen T, Sagvolden G, Pran K, Olsen T (2016) Bond strength and performance of optical fibre Bragg gratings sensors embedded in composite patch repairs for military aircraft
27. Mandell JF, Chen J-H, McGarry FJ (1980) A microdebonding test for in-situ Fiber-Matrix Bond Strength and Moisture Effects, MIT Research Report 80-1
28. Grande DH, Mandell JF, Hong KCC (1988) Fibre-matrix bond strength studies of glass, ceramic, and metal matrix composites. J Mater Sci 23:311–328
29. Ji X, Dai Y, Zheng B-L, Ye L, Mai Y-W (2003) Interface end theory and re-evaluation in interfacial strength test methods. Compos Interfaces 10(6):567–580. doi:10.1163/156855403322667278
30. Shen CH, Springer GS (1976) Moisture absorption and desorption of composite materials. J Compos Mater 10(1):2–20
31. Fick A (1855) Ueber diffusion. Ann Phys 170(1):59–86
32. Woo M, Piggott MR (1988) Water absorption of resins and composites: IV. Water transport in fiber reinforced plastics. J Compos Technol Res 10(1):20–24
33. Mallarino S, Chailan JF, Vernet JL (2005) Interphase investigation in glass fibre composites by micro-thermal analysis. Compos A 36:1300

34. Joliff Y, Belec L, Chailan JF (2013) Modified water diffusion kinetics in an unidirectional glass/fibre composite due to the interphase area: experimental, analytical and numerical approach. Compos Struct 97:296–303
35. Joliff Y, Rekik W, Belec L, Chailan JF (2014) Study of the moisture/stress effects on glass fibre/epoxy composite and the impact of the interphase area. Compos Struct 108:876–885
36. Gagani A, Echtermeyer AT, Anisotropic diffusion in glass epoxy laminates, to be published
37. Perreux D, Choqueuse D, Davies P (2002) Anomalies in moisture absorption of glass fibre reinforced epoxy tubes. Compos Part: A Appl Sci Manuf 33(2):147–154
38. Davies P, Baizeau R, Choqueuse D, Salmon L, Nagot F (2000) Aging and long term behaviour of composite tubes. Proc Duracosys 99:143–151
39. Saeb S, Staenman P, Javili A (2016) Aspects of computational homogenization at finite deformations: a unifying review from reuss' to voigt's bound. Appl Mech Rev (2016)
40. Otero F, Ollez S, Martinez X, Salomon O (2015) Numerical homogenization for composite materials analysis. Comparison with other micro mechanical formulations. Compos Struct
41. Geers MGD, Kuznetsova VG, Brekelmans WAM (2010) Multi-scale computational homogenization: trends and challenges. J Comput Appl Math
42. Talreja R (2006) Multi-scale modeling in damage mechanics of composite materials. J Mater Sci (2006)

Present and Future Composites Requirements for the Offshore Oil and Gas Industry

Denis Melot

Abstract The aim of this paper is to review some of the main applications of composite and polymer materials in the oil and gas industry. It starts with the basics of material selection and the methods used for corrosion mitigation of carbon steel which is the base case and material of choice due to its low cost and ease of installation. Some specific business cases are identified for composite materials and some typical applications are described in more details.

Keywords Corrosion-composite-polymer-oil · Gas-offshore-flowline · Oil and Gas

1 Introduction

The oil and gas industry is a "metal-based" industry, and more specifically, a carbon steel-based industry.

At the same time, polymer and composites materials are recognized and used in the industry for the general following reasons:

- Corrosion resistance (maintenance-free).
- Better strength to weight ratio than steel.
- Light weight/ease of installation.

In order to switch from well-known solutions with associated codes and standards (like for metallic solutions) to alternative materials like composites or polymers, strong business cases must be established.

D. Melot (✉)
TOTAL SA, Courbevoie, France
e-mail: denis.melot@total.com

P. Davies and Y.D.S. Rajapakse (eds.), *Durability of Composites in a Marine Environment 2*, Solid Mechanics and Its Applications 245,
https://doi.org/10.1007/978-3-319-65145-3_9

Some applications are already well known such as (and not limited to) the following:

- Flexible unbonded pipes, with the advantage of flexibility versus rigid steel pipe leading to ease of installation (reeled pipe/no detailed metrology/cope with uneven topography/work under dynamic conditions…) [1].
- Fiber glass reinforced thermoset pipes (typically referred to as GRE: Glass Reinforced Epoxy or GRP: Glass Reinforced Plastic), with the advantage of no corrosion/no maintenance/no cathodic protection and ease of installation for offshore piping or onshore flowlines.
- Composite gratings for offshore application, with the advantage of no corrosion and light weight.
- Sea water piping at moderate pressures ($\sim$15–30 bars).

New product developments are also being brought to the market, either to meet the increasing demands of offshore developments as described by Groves [2] or to push the limits of existing solutions such as, for flexible pipes:

- Anti-H_2S layer to allow slightly sour fluid transportation and use in ultra deep water [3].
- Cost effective polymer qualification as pressure sheath [4].
- Higher temperature range mono-layer pressure sheath [5].

In addition, new manufacturing techniques lead to the recent appearance on the market of the so-called TCP (Thermoplastic Composite Pipes), which are spoolable, fully bonded composite pipes, and which close the gap between the rigid classical GRE pipe and the flexible unbonded pipes [6]. Though spoolable, these pipes are not yet as flexible as unbonded pipes but are cheaper to manufacture with higher pressure and combined loadings capabilities than the GRE pipes.

One existing identified business case for TCP is its use as downline for pre-commissioning of pipelines or well intervention [7, 8].

End users have been reported to use composites and polymers for their production facilities. Petrobras states that large cost reductions were obtained by changing out cable trays, gratings, piping, and chemical storage tanks from steel to composites on their offshore facilities where weight reduction and sea water corrosion resistance are critical. As an example, they claim that since 1994, all their new platforms have composite gratings and that more than 40% of the steel gratings are replaced by composites [9].

Melot et al. [10] list the use of polymer materials within their upstream facilities. They note that the material of choice remains carbon steel and that the development of composite/polymer parts is foreseen as fields become mature producing more water at a higher temperature with increased corrosivity.

Paiva [11] notes the need for higher temperature and higher pressure resistant materials (either metallic or polymers). The higher corrosivity of the produced fluids is also noted, which enhances the need for higher corrosion resistant materials or coatings.

Qi [12] mentions that over 17,000 km of composite pipes have been installed by CNPC (one local oil and gas company) and it is growing at about 2000 km per year. The advantages noted are similar to those presented here, with a focus on corrosion resistance for production fluids with high amounts of acid gases (H_2S and CO_2). The low life cycle cost of composite pipes is also highlighted.

Melve et al. [13] present a new technology qualification process for a composite kill and choke line. It is a hybrid structure made of a steel liner reinforced by filament wound layers of carbon fiber-epoxy resin and is designed for a 15000 psi internal pressure with an internal diameter of 4.5 in. Its use is for deep offshore applications down to 4000 m depth to reduce the overall weight of the drilling riser.

All these examples show that the advantages of polymer materials are well identified within the industry, but that their use is linked to specific business cases. Yet there is a consensus in the industry that the use of alternative materials to carbon steel will grow in the future due to pressure to lower life cycle costs and reduce weight for ease of installation or for ultra deep offshore developments.

2 Material Selection in the Oil and Gas Industry

The fluid coming out of the production wells is a three phase fluid (i.e., hydrocarbons/water/gas in different proportions) under pressure and temperature.

The fluid processing mainly consists in the separation of these three phases followed by a specific treatment for each of the phases (see Fig. 1).

The water produced can be reinjected in the field to improve oil recovery. In a similar way, part of the produced gas can be used to help the produced fluid to reach the process facilities (gas lift process).

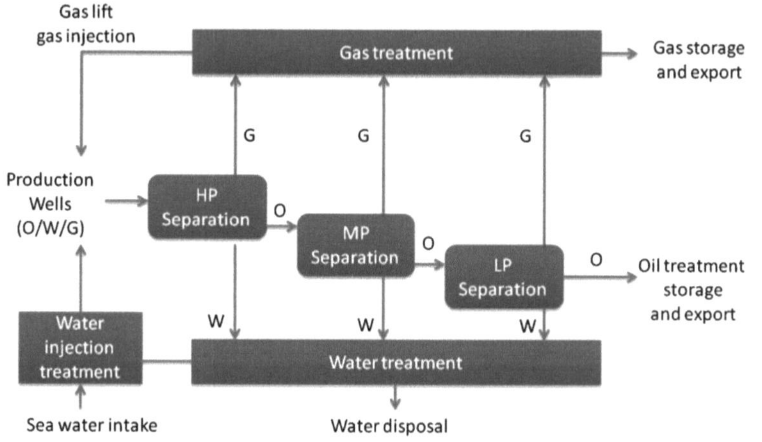

Fig. 1 Schematic of a typical fluid process

2.1 A Corrosion Assessment in Oil and Gas

The most frequent cases of corrosion in the oil and gas industry are as follows:

- H_2S cracking that may happen within days.
- CO_2 corrosion and microbial induced corrosion that happens within years.
- Atmospheric corrosion in the marine environment that happens within 10 years.

It should be mentioned here that in the produced fluids, oxygen is not present. Corrosion of carbon steel equipment is happening via acid gases (CO_2 and H_2S) dissolved in the water phase.

In that respect, the pH of the water phase in a produced fluid is one of the key parameters to assess the corrosivity of the fluid. Depending on the water composition, the amount of CO_2 and H_2S and the total pressure, this pH may typically vary from 3 to 6.5. The limited water wetting of the carbon steel and the possible protection from corrosion layers may also play a mitigating role in the corrosion process.

The majority of the corrosion concern is coming from CO_2 corrosion (even if H_2S cracking is much more critical in terms of risk). As the objective of this paper is to give only some general features of the material selection and corrosion assessment, we will focus mainly on this type of corrosion as a typical example.

The CO_2 corrosiveness of production fluids can be evaluated using predictive models.

Several predictive models have successively been issued since the 70's, both from research entities and from operators. The oldest and best known is the very simple "De Waard–Milliams" model [14, 15], which has constituted the "matrix" of the more elaborate current models. Yet there are, at present, no generally accepted prediction models in the industry and almost every company has developed its own model. Among those, operators tend to adjust their model to their own field experience and their practical needs.

TOTAL has developed its own model over many years, initially called Corplus, recently renamed PreCorr since its last major up-date. It includes all physico-chemical equilibria involved in CO_2 and H_2S corrosions, the kinetics of corrosion, thermodynamics and hydrodynamics, water/oil wetting, and protection performance of corrosion layers. From its early days, a particular emphasis has been given to the necessary inclusion of the whole field experience of the Company, through semi-empirical classification criteria of the Corrosiveness.

It provides the corrosion engineer with a "CO_2 corrosiveness" of the produced fluid with respect to a carbon steel surface, which can then be translated into a "corrosion likelihood" of the equipment subjected to this fluid (i.e., the ability of equipment to sustain a fluid over a certain period). This can be translated into different categories as reported in Fig. 2 below (Fig. 3, 4).

A few typical examples of CO_2 corrosion are given in the pictures below.

	Typical CO_2 Corrosiveness for CS (mm/y)	Approximate lifetime vs design life (assuming no preventive action) - CO_2 Corrosion Likelihood -	Practical likelihood of failure
Very Low	*< 0.1*	*System will last much longer than required with no failures, without any preventive action*	*None*
Low	*0.1 – 0.3*	*System should reach its initial design life without a failure with available corrosion allowances, not requiring any complementary preventive action*	*Low but no extension over initial design life*
Medium	*0.3 to 1*	*System should reach 50% of its design life before a failure occurs, if no preventive action is undertaken*	*Likely but delayed*
High	*1 to 3*	*A failure may occur after 25% of its design life, if no preventive action is undertaken*	*Very likely within a few years*
Very High	*> 3- 5*	*The system may not even resist ¼ of its design life, if no preventive action is undertaken*	*Certain and soon*

Fig. 2 Corrosion likelihood assessment as per TOTAL model

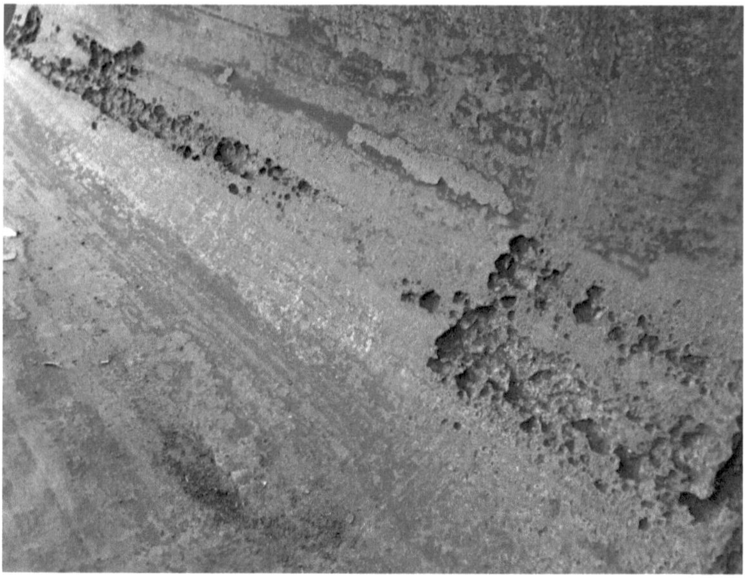

Fig. 3 CO_2 internal corrosion on carbon steel pipe

Fig. 4 CO_2 internal corrosion on carbon steel pipe

2.2 *Corrosion Mitigation Methods*

Once the corrosion likelihood is known, mitigation measures can be implemented to keep the use of carbon steel safe and reliable.

The first and easiest mitigation measure is the implementation of a corrosion allowance to the equipment. A few mm up to 10 mm of extra thickness can then be prescribed at the design phase of the equipment that will take care of the corrosion rate without any additional treatment. However, this measure in itself is rarely sufficient to ensure complete corrosion mitigation over 20 years or more, except for very moderate corrosiveness. It is rather aimed at offering some margin in the case of a temporary lack of efficiency of the other mitigation solutions mentioned below.

In addition to the above, fluid inhibition by specialized chemicals at low dosage is the conventional solution in the industry for corrosion mitigation on facilities built with carbon steel.

The most common corrosion inhibitor treatment consists of a continuous injection at the inlet of the equipment to be protected, at typical dosages of 10 to 100 ppm (v/v) versus the liquid or water flow rate.

pH stabilization can also be applied to wet gas/condensate pipelines (using chemicals such as MDEA, KOH …) where the condensed water has a particularly low pH and is highly corrosive.

When it comes to the water injection network, the fluid can be a mix of produced water (separated from the oil and gas phase in the process-see Fig. 1) and sea (or lake) water, in highly variable proportions. Its corrosivity is usually mitigated via the following measures:

- Oxygen removal, for the part of the water coming from the outside (sea or lake),
- Biocide injection to prevent the Microbiologically Influenced Corrosion,
- Corrosion inhibitor treatment, to deal with the acidity coming from the produced water, if any,
- Possibly cleaning pigs (for deposit removal).

The corrosivity of the gas network (see Fig. 1) is associated with its condensed water, which is the most acidic of all produced waters (pH usually within 3 to 4.5). As such, the preferred mitigation measure is to dry the gas by a dedicated dehydration process, to ensure that its dew point is always below the minimal temperature that the gas may reach. However, as this dehydration is only done in the final processing part of production facilities, corrosion inhibitors are also used on the gas side, upstream of this dehydration process.

As noted previously, the objective of all these mitigation measures is to keep the use of carbon steel as widespread as possible by controlling its corrosion rate. This is generally done successfully as, in general, carbon steel represents about 95% of the steel consumption in the oil and gas industry. Nevertheless, as discussed below, all these measures may face practical limits of application or pose too serious cost or reliability issues to be applied in numerous cases.

2.3 *Corrosion Monitoring*

Whenever a corrosion mitigation measure is implemented, there is a requirement to control its efficiency throughout the lifetime of the equipment.

A dedicated field of expertise exists nowadays in the industry. The most common monitoring methods implemented are as follows:

- Follow-up of chemical injection specifications.
- Monitoring of dehydration process (for gas treatment).
- Monitoring and maintenance of oxygen removal process (for water treatment).
- Corrosion probes and coupons downstream of chemical injection.
- Periodic wall thickness checks and periodic inspection.
- Miscellaneous chemical analyses.

- Bioprobes (for microbial induced corrosion).
- Bacteria count in water.

The oil and gas industry has learned to cope with carbon steel corrosion risks and most important to keep corrosion rates under control and the failure rate at an acceptable level.

2.4 Practical Issues with Mitigation Measures

There are cases identified where the mitigation measures described above are not efficient or cost effective or considered high risk. These are for instance as the following:

- Production wells, where the chemical injection is often not reliable and costly, specially, offshore, as the injection must be done at the bottom of the wells, i.e., some thousands of meters below the surface.
- Subsea production manifolds, jumpers, Christmas trees, where a high reliability is a must but is very difficult to achieve.
- Topsides for instrument tubes, chemical/hydraulic lines.
- Produced water networks.
- Topsides for open and closed drains where a mix of unknown aerated fluids are commonly leading to fast and multiple losses of containments.

There are also potential issues associated with corrosion inhibitors themselves as follows:

- Their efficiency as mitigation measure is quite specific and only proven after a few years of use in the field. It can also evolve as fluid composition and temperature change with time.
- Their injection and dosage on-site may not always be optimum and need a stringent follow-up (good condition of pumps, daily verification of injection, weekly verification of injected volumes, etc.)
- Solid particles and scale accumulation may compromise their efficiency.
- They become hardly efficient above 110–120 °C.
- Their compatibility with other added chemicals can be problematic.

2.5 The use of Alternative Materials to Carbon Steel

Whenever inhibition is not safe enough or inefficient or not cost effective for some of the above reasons, the natural alternative choice in the oil and gas industry is to move toward a corrosion resistant alloy (CRA) such as the following:

- 13% Cr alloy for well tubing.
- Duplex or super duplex for subsea equipment, wet gas processing, instrument tubing…

This, of course, comes with a cost (around 5–10 times higher compared to carbon steel pipe or piping).

It is in this window that non-metallics must also find their place, as long as they can be as attractive as or even more attractive than both carbon steel + mitigation and CRAs.

2.6 Conclusions

The use of metallic materials (either carbon steel or corrosion resistant alloy) is linked to the following:

- Availability and cost.
- Ease of installation.
- Well-known mitigation measures.
- Available inspection and monitoring techniques.
- Good knowledge of failure modes and aging process (i.e., corrosion).

The use of polymer materials as a means of internal corrosion mitigation must overcome all the above items and must be linked to a specific business case.

In addition, the use of corrosion resistant alloys is limited in quantity and may end up being limited in their availability in the future. This also means that the use of polymer/composite materials is linked to additional benefits such as the following:

- Life cycle cost savings (specially, versus corrosion resistant alloys).
- Ease of installation (light materials/quick connections/light laying equipment), versus all metallic materials.
- Weight saving (for offshore structures and ultra deep offshore developments), also versus any metallic material.

3 Identified Potential Business Cases for Polymer/Composite Materials

Some potential business cases have been identified in the industry for the use of polymer/composite materials as the following:

For internal corrosion:

- Transport of nontreated sea water (water intake/fire fighting).
- Mature fields with high water cut (above 80% typically) where a proper fluid inhibition and reliability is costly and problematic.
- Aerated and chlorinated water line offshore.
- Water disposal wells (onshore and offshore).
- Aquifer water injection lines (with typically 300 mg/l NaCl).

For external corrosion and light weight/ease of installation:

- Gratings for offshore splash zone area.
- Secondary structures in offshore where weight is critical such as Tension Leg Platforms.
- Composite risers for ultra deep offshore.
- Flexible and/or composite spoolable pipes for ease of installation, especially subsea (no welds).

These above are just a few general examples and some others might exist.

What is critical is to get into the details of each business case in order to clearly highlight the interest of composite materials and challenge as much as possible the use of metallic materials whenever possible. As such, not only the fluid corrosiveness or the operating conditions (P/T) are critical, but also the installation process and the life cycle cost including the use of chemicals. This is of course not always easy, especially because in our industry the CAPEX (Capital Expenditure) part is much more critical than the OPEX (Operating Expenditure) part.

In the following sub-chapters, we will detail the different polymer/composite solutions that are available to the oil and gas engineer.

3.1 Glass Fiber Reinforced Epoxy Pipes

Composite pipes are a combination of resin (matrix) and long fibers. The design of such pipes is critical to the application:

- Orientation and type of fibers give the expected mechanical and chemical compatibility (type C glass fibers are more chemically resistant than type E glass fibers).
- Chemistry of the resin gives the temperature rating and overall chemical resistance of the pipe as the following:
  - Polyesters are used for large diameter pipe to convey water or sewage.
  - Vinylesters are used for more aggressive fluids (like strong acids for example).
  - Epoxies are used for aggressive fluids and high-pressure applications.

The ISO 14692 standard defines the general temperature ratings of the pipes according to the chemical nature of the matrix as follows:

- 110 °C for epoxy.
- 100 °C for vinylester.
- 70 °C for polyester.
- 150 °C for phenolic.

Some manufacturers also define the temperature limitations of the epoxy matrix according to the chemical nature of the hardener:

- 65 °C for anhydride type epoxy.
- 93 °C for amine type epoxy.
- Up to 200 °C for aromatic type epoxy.

Of course, these are all related to the expected glass transition temperature of the matrix and we consider as a base case that the operating temperature of the pipe should not be above the glass transition temperature of minus 20 °C.

The overall structure of a composite pipe is as follow:

- Internal layer of approximately 0.5 mm composed mainly of resin for improved chemical resistance.
- Structural wall for the mechanical requirement.
- Top coat of approximately 0.3 mm for UV resistance.

The internal pressure rating for composite pipe is calculated via long term (ASTM D2992) and short term (ASTM D1599) testing results.

The joining system is critical to the integrity and pressure rating of the lines. It can be adhesively bonded or threaded depending on the pressure rating and/or the installation constraints. To learn more on this specific subject, please refer to ISO 14692, API 15LR, and API 15 h.

Fiber glass reinforced epoxy pipes have been used by TOTAL since the early 80s in France on mature fields where the amount of water produced is typically above 80%. The driving force to switch from carbon steel to composite pipes was clearly internal corrosion mitigation. The operating conditions were also favorable with low pressure (about 10 bars) and mild temperature (40–80 °C). In 1983, 67 km of pipes were installed and in 1993, 130 km were in operation with no issues. For inspection purposes, pipe samples are recovered from the lines at regular intervals (every 5 years) and pressure tested to burst. So far, no test results have shown any degradation in the pressure rating compared to the original pipe sample. The pictures below, Figs. 5 and 6, illustrate some typical applications of composite pipes onshore.

Composite pipes are also used for the fire fighting systems on offshore facilities again for internal corrosion mitigation (untreated stagnant sea water). Pressure rating is low (18b) and joining system is the "bell and spigot" type. An example of this application is given in the following figure (Fig. 7).

The main issues related to this application are as follows:

Fig. 5 Composite pipe installation for flowlines (Courtesy of SBPI)

- Water hammer effect leading to debonding and cracking at joint interfaces
- Joining defects.
- Vibrations.
- Bad design (especially for supporting fixtures).

This leads to early failures, most of the time identified during commissioning of the lines. After proper repair of the defective joints, the lines operate without any identified issues. Yet, these early failures do not play in favor of the use of composite pipes and make operators less confident with these solutions.

Another identified business case is the use of such pipes for process open and closed drains network, where the fluids conveyed are of unknown origin and where oxygen entry is common, leading to a very high fluid corrosivity which is not possible to mitigate via the usual measures (inhibition is inefficient). In the specific

Fig. 6 Composite pipes for water transportation (Courtesy of SBPI)

case of closed drain systems, extra care is needed to ensure that the pressure rating of the whole network is considered in the design of the composite network.

These composite pipes are not used for subsea application due to their limited mechanical properties, and low collapse resistance.

3.2 Spoolable Reinforced Thermoplastic Pipes

Glass fiber reinforced epoxy pipes have a connection every 9 m and have very limited minimum bending radius. By making the composite pipe more flexible and even spoolable, the cost savings in installation time, pipe transportation and lower risk of failure due to limited numbers of joints are significant. This has led several pipe suppliers to provide the industry with such alternative products.

There are as many reinforced pipe designs as there are manufacturers. Most of the time, they are based on the general structure (Fig. 8).

Depending upon the operating temperature and type of fluid of the line, polyethylene can be changed to a more suitable thermoplastic such as polyamide or PVDF. The reinforcing layers can be made of aramid fibers, steel strips, dry glass fibers or even glass fiber reinforced epoxy matrix, all being bonded or not to the liner.

Fig. 7 Fire fighting system network on an offshore facility (Courtesy of SBPI)

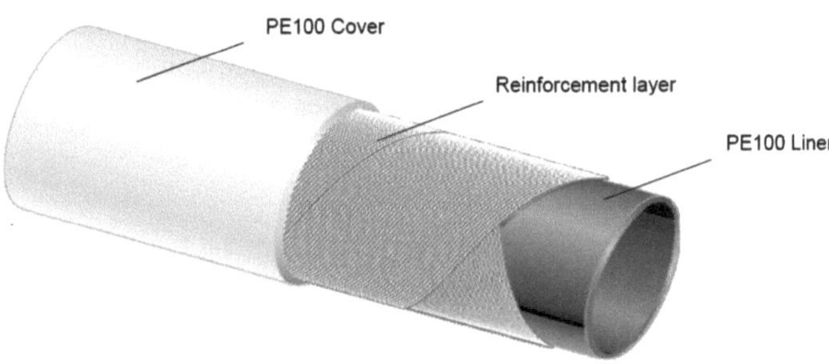

Fig. 8 Typical structure of a spoolable reinforced thermoplastic pipe

The length of a single spool ranges from 120 up to 2000 m depending of course upon the diameter of the line. For a 4 in. line, the typical spool length goes from 500 to 700 m. The pipe diameter is limited to 8 in. maximum at present.

The qualification process follows the requirements of the API15S, the pressure rating is established via the regression curve of ASTM D2992 (similar to those described in Sect. 3.1 and polyethylene pipes). There is no requirement for axial loading and the bending capability is tested only via 1000 h survival pressure testing on bent samples (and after 10 reverse bending cycles). External pressure is tested via compression ring test only.

This makes these type of pipes applicable mainly to onshore networks with no complex load cases. A few examples exist of use in shallow water.

Typical examples are water injection lines and industrial water distribution onshore. They have also been used for unburied production fluid gathering lines in the Middle East.

The contact with production fluids is limited by swelling and diffusion properties of the thermoplastic liner used. New products are being developed to mitigate this issue via the introduction of a metallic layer within the pipe structure to make it gas and fluid tight [16].

3.3 Thermoplastic Composite Pipes

Thermoplastic composite pipes are a new development of the spoolable reinforced thermoplastic pipes described in Sect. 3.2 with a very similar structure (see Fig. 9).

All the layers are fused together which makes these pipes a solid wall structure.

They can accommodate high-pressure ratings and complex loading cases and can be made collapse resistant unlike the pipes in Sect. 3.2.

As such, they can compete in performance with carbon steel pipes or unbonded flexible pipes from a purely structural point of view.

The general qualification of these pipes follows the requirements of the DNV RP F119 standard. The philosophy of this standard is to perform extensive testing at the laminate level to be able to build up a predictive model for the pipe behavior and limit the full-scale testing on the pipe to the validation of the model. Complex load cases can be taken into account and predicted. Each application or business case leads to a specific pipe design (number of reinforcement layers/angle of the fibers in each layer, etc.) made to fall within the acceptance criteria and safety factors of DNV RP F119.

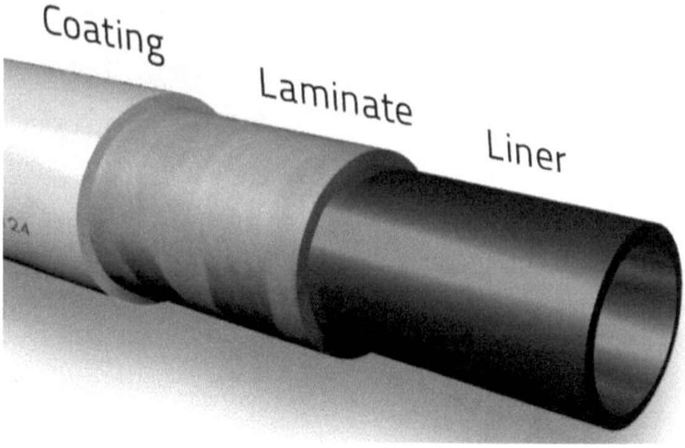

Fig. 9 Structure of a thermoplastic composite pipe

Due to their higher overall performances, these pipes have limited flexibility compared to spoolable reinforced thermoplastic pipes of Sect. 3.2 or unbonded flexible pipes, but are still spoolable to a minimum bending radius of 2–7 m depending on the structure and diameter. The diameter range is limited to 6–8 in. at present. Each spool can be as long as 3000 m.

One interesting feature of some pipe designs is the possibility to mount the end fitting directly on-site during installation by cutting at length the spooled pipe. This makes it easier for rehabilitation/retrofit on existing installations and avoids the need for precise metrology in advance. A few spools could be ordered and used by the meter on demand.

The downlines for well intervention or pipe commissioning are one existing application of these pipes [7, 8] (Fig. 10). Here the main features are the light

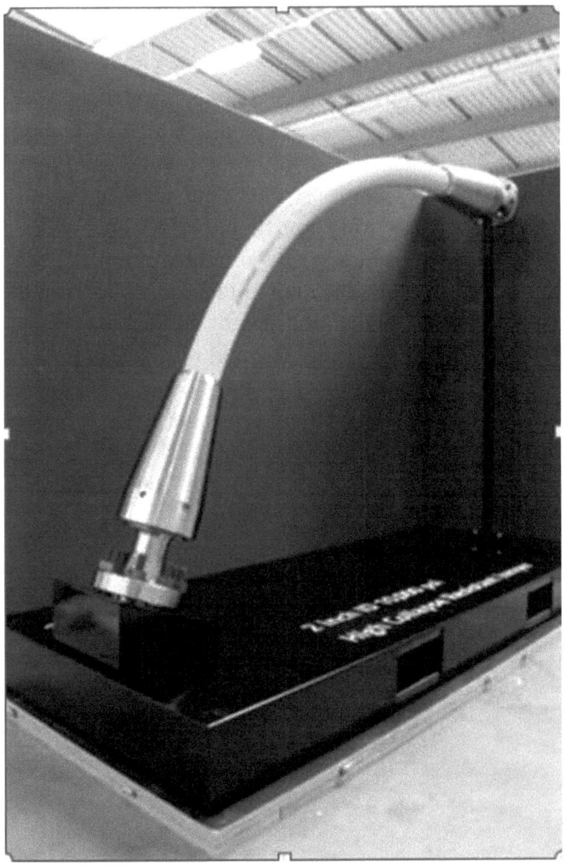

Fig. 10 Thermoplastic composite pipe (Courtesy of Airborne Oil and Gas)

Fig. 11 Spooled Composite line downhole deployment (Courtesy of Airborne Oil and Gas)

weight of the pipe (meaning lighter vessels and lifting equipments for installation) and better resistance to the cycle of reeling–unreeling compared to carbon steel lines. The aging process and monitoring of the pipe condition are made easier due to the retrievablility of the line between two campaigns (possibility to inspect and take samples of the pipe for testing) (Fig. 11).

At the moment, further qualifications are ongoing for the use of such products for permanent subsea water injection jumpers where ease of installation is a key factor. Chevron North Sea Ltd. installed in 2016 a thermoplastic composite methanol injection spool subsea for permanent use.

Further developments are foreseen for their use as risers for ultra deep water (4000 m and more) in the coming years. The knowledge of the aging process of the product in contact with production fluids and its monitoring with time will be a key issue for the development of such business cases.

3.4 Internally Lined Carbon Steel Pipes

Even if slightly out of the scope of this paper, the use of thermoplastic liners to protect the internal surface of a carbon steel pipe from the conveyed fluid must be discussed and presented here, as it is a potential competitor (already used in the industry) to composite pipes.

The general principle is to insert a thermoplastic pipe into the carbon steel pipe by pulling it through a diameter reducing box. Once in the carbon steel pipe, the liner is released from its tension and allowed to freely relax back. It ends up tight fitted to the carbon steel pipe [17].

The maximum length of a liner that can be inserted in one go is dependent on its mechanical properties in tension but the average length can be estimated between 1500 and 2500 m. The connections between two strings of lined pipes are made of stainless steel at present, to ensure the continuity of the anti-corrosion barrier through the pipe. New connection designs under qualification use thermoplastic sleeves electro-fused to the liner on-site for a faster, easier and cheaper installation cost of the technology.

The liner technology was first implemented in the North Sea for offshore water injection lines in 1994. The liner was polyethylene with stainless steel connections. Since then several hundreds of kilometers of lined pipes have been installed in the world.

One of the main advantages of the liner technology is that it uses the carbon steel pipe as structural support.

One of the key qualification issues for the use of liners for gas and hydrocarbon transportation is the knowledge of the diffusion and permeability properties of the liner to production fluids and gases to evaluate the risk of collapse during operation due to rapid decompression, and assessment of the composition of the annulus (space between the steel and the liner) and the quantity of gas trapped in the annulus (there may be a need for venting to release the pressure), see Fig. 12.

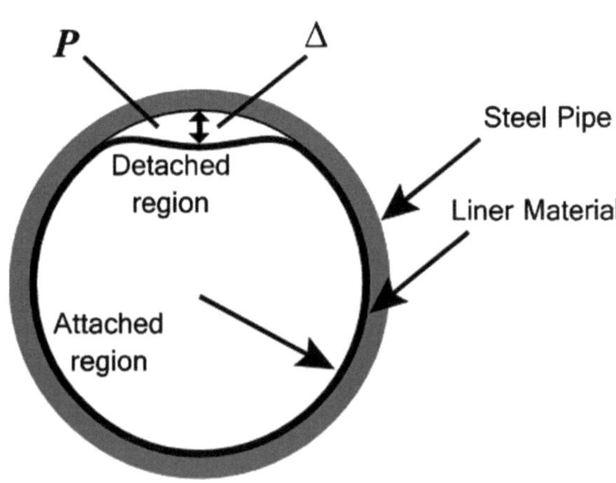

Fig. 12 Schematic diagram of the collapse mechanism. The pressure build up in the detached region is coming from the species that have permeated through the liner. Collapse is resulting from a depressurization in the line during operation

3.5 Composite Gratings

When looking at external corrosion mitigation, one typical example, even if not the most technologically challenging, is the use of composite gratings for offshore structures.

The common practice in the industry is the use of standardized galvanized gratings. In offshore environments, especially in the splash zone areas of offshore structures, such gratings can have a limited lifetime of about 5 years.

Composite gratings can be either molded (65% resin, isotropic, cheap) or pultruded (70% fibers, anisotropic, high mechanical resistance).

The most commonly used resins are as follows:

- Isophthalic polyester for good overall properties and low cost.
- Vinylester for improved chemical and mechanical resistance.
- Phenolic for fire resistance and mechanical resistance.

The oil and gas industry uses mostly pultruded phenolic type gratings for fire resistance performance and mechanical properties.

Molded gratings are used whenever fire resistance is not required or when high load bearing resistance is not necessary.

Typical examples of composite gratings offshore are given in the following pictures, Figs. 13, 14, and 15.

Apart from the corrosion resistance aspect, weight saving is also a good business case for composite gratings as they are typically 50–70% lighter than steel gratings. This can be critical on some offshore structures such as tension leg platforms.

Fig. 13 Pultruded phenolic grating on offshore structure

Fig. 14 Molded grating on offshore structure

Fig. 15 Composite grating in the splash zone area of an offshore structure

4 Conclusion

The oil and gas industry is a carbon steel-based industry. The corrosion is mitigated via fluid treatment and chemical injection for internal corrosion and cathodic protection and coating for external corrosion. The steel aging process (corrosion) and failure modes are well known and can be monitored and inspected in most cases.

The use of polymer and composite materials must be linked to detailed and specific business cases where corrosion mitigation is not possible or recognized as unreliable with high associated corrosivity.

Polymer and composite materials are not used up to their full potential in the oil and gas industry due to the following:

- Lack of industry knowledge and lack of detailed business cases.
- Lack of simple qualification process and non standardized solutions.
- Unknown failure modes and difficulty to extrapolate the aging to the end of life of the equipment (the Arrhenius law usually does not work).
- Lack of qualified contractors for installation on-site.
- Lack of inspection techniques for installation and monitoring during operation.

Nevertheless, their intrinsic characteristics make them the solution of choice for the future of the industry for maintenance-free and easy-to-install solutions.

Some typical examples are reported in this paper including new technologies such as thermoplastic composite pipes that show the potential for a wider use of these solutions.

The current oil and gas market, calling for cost reduction in project development with innovative solutions, is certainly a great opportunity for composites and polymers.

References

1. Cox D (2010) Flexible pipe: the story so far. In: Asia offshore engineering conference, 30th Sept. 2010
2. Groves S (2006), Meeting the increased demands of offshore developments. Oilfield engineering with polymers
3. Chirat M et al (2015) Anti-H2S layer for flexible pipe-full scale flowline qualification tests. In: Offshore technology conference 2016, OTC-25775-MS
4. Tronc F et al (2016) Qualification of a new robust and cost effective polymer for flexible pipe pressure sheath. Offshore technology conference Asia 2016, OTC-26056
5. Epstein T et al (2016) Taking further the domain of use of low plasticized high molecular weight PVDF for monolayer dynamic application in flexible pipe. In: 35th Conference on ocean offshore and artic engineering, OMAE 2016–54650
6. Steuten B (2016) Reduce project and lifecycle cost with TCP flowlines, Offshore technology conference Asia 2016, OTC-26512
7. Van Onna et al (2014) Evaluation of the first deployment of a composite downline in deepwater Brazil. Rio oil and gas conference 2014, IBP 1852-14

8. Kremers et al (2015) Thermoplastic composite pipe technology: development qualification and building track record. Compos Adv Mater Expo
9. Pesoa GA et al (2001) Petrobras composite topside applications3. Compos. Mater. Offshore Appl CEAC 2001:437–448
10. Melot D et al (2009) The use of polymeric materials with Total upstream. In: Eurocorr conference
11. Paiva A (2013) Challenges and opportunities for materials in the oil and gas industry. In: High performance polymers for oil and gas applications
12. Qi D (2010) Applications of polymer composite pipes in oilfield in China. Oilfield Eng Polymers 2010
13. Melve B et al, (2014) Composite material choke and kill lines-experience from qualification and installation, Proc. Oilfield Engineering with Polymers conference, AMI London
14. Waard De et al (1991) Predictive model for CO_2 corrosion engineering in wet natural gas pipelines. Corrosion 47(12):976
15. De Waard et al (1995) Influence of liquid flow velocity on CO_2 corrosion-a semi empirical model. Corrosion, Paper 128
16. Dalmonen B (2016) mitigation of permeation issues in spoolable reinforced plastic linepipes. Proc. Oilfield Engineering with Polymers conference, AMI London
17. Feeney A (2014) Polymer lining technology. Proc. Oilfield Engineering with Polymers conference, AMI London

Composite Materials in Tidal Energy Blades

Matthew Dawson, Peter Davies, Paul Harper and Simon Wilkinson

Abstract Composite materials are the natural choice for the construction of tidal energy blades; their high strength, good environmental resistance and the ease with which they may be used to form complex shapes means that they are well suited to the application. The design of MW-scale tidal energy blades has evolved over a number of years, keeping step with the requirements of turbine developers as their emphasis shifts from prototyping to production. In parallel, it has been necessary to develop a rigorous approach to materials testing and qualification. The specifics of the blade structure and the operational environment mean that it has been necessary to solve a number of complications that arise during such a qualification programme. These issues are discussed and efforts to mitigate their consequences are explained. Although this chapter is written particularly with tidal energy blades in mind, much of the content will be applicable to a wide range of other applications where composite materials are used in the marine environment.

1 Introduction & Background

Tidal energy blades (Fig. 1) represent a relatively new and challenging application for composite materials. They present a combination of commercial, environmental and loading conditions that demand a detailed engineering approach. In particular, it is necessary for a cost-effective structure to sustain very heavy fatigue loading in a submerged marine environment.

M. Dawson (✉)
AEL/Airborne International, Marine BU, Lambourn Woodlands RG17 7TJ, UK
e-mail: matthew.dawson6@btinternet.com

P. Davies
IFREMER Centre Bretagne, 29280 Plouzané, France
e-mail: peter.davies@ifremer.fr

P. Harper · S. Wilkinson
ACCIS, University of Bristol, Bristol BS8 1TR, UK
e-mail: paul.harper@bristol.ac.uk

P. Davies and Y.D.S. Rajapakse (eds.), *Durability of Composites in a Marine Environment 2*, Solid Mechanics and Its Applications 245,
https://doi.org/10.1007/978-3-319-65145-3_10

Fig. 1 Various tidal energy blades in situ, manufactured by AEL/Airborne

Over the past decade, tidal turbines have been built to a wide variety of designs. Whilst the wind industry has converged on a 3-bladed horizontal axis turbine as the preferred solution, the tidal industry is still relatively young and there is not yet a consensus regarding the optimal configuration for a tidal turbine. In particular, developers are unsure whether it is more cost-effective to opt for a complex device that generates as much energy as possible or a simpler device that is relatively cheap to manufacture and maintain but is unable to extract as much energy from a given flow. It follows that the argument has not been won or lost with regard to a number of key design variables, and progress for many developers has been hampered by the inconsistency of government and industrial support, often meaning that operational experience is still fairly limited. Inspection of existing designs and prototypes would lead to the conclusion that designers have a number of different ideas with regard to the following aspects, amongst others:-

- *Scale*:-
  Generating capacity varies from several kW to more than 1 MW. Some developers look for economies of scale in terms of production quantities, whilst others seek it in terms of machine size.
- ***Foundation***:-
  Machines exist whereby turbines are attached to fixed (gravity base or piled), floating and even 'flying' (underwater) support structure.
- ***Method of deployment/retrieval***:-
  Costs associated with deployment and retrieval are very significant, so methods vary greatly as designers strive to minimise the overall costs of these activities.
- ***Blade mounting***:-
  Blades may be mounted on a central hub or an outer ring. Furthermore, some designers favour variable pitch blades in a drive to maximise energy capture and minimise loads, whilst others prefer a simpler machine, and therefore opt for a fixed-pitch scheme. The phrase *"if it hasn't got it, it can't go wrong"* is particularly apt here. The cost of rectifying problems/breakdowns in service is very high, so reliability is extremely important.
- ***Number of blades per rotor***:—This may vary from 2 to 10.

In many cases there are strong interdependencies between the varying design options, meaning that identification of any particular 'best' answer is far from straightforward (in any case the optimal solution may be site-dependent). It is likely that the field will narrow in time, as weaker solutions and developers fall by the wayside.

Whilst there are clearly many permutations in terms of the configuration of the overall turbine, it is fortunate for the blade designer that many of these options do not especially affect the blade concept design. However, the selection of a fixed or variable pitch system would have implications for the favoured blade concept. A variable pitch blade will tend to terminate in a circular root to efficiently mate with the pitch bearing; this requirement in itself has design implications that cascade along the rest of the blade. However, a fixed-pitch blade typically has no need for a circular root, so it gives the blade designer more options in terms of configuration. Unsurprisingly, the decision to mount the blades at the hub or to an outer ring is also very significant in terms of blade concept selection.

A variety of materials and configurations has also been adopted for the construction of tidal blades, depending on the specifics of the application and the experience and preferences of the designer. It is generally the case that blades are built with epoxy matrix composite materials, incorporating both carbon and glass reinforcements as appropriate. Whilst glass fibres are significantly cheaper and, therefore, represent the default choice, there is also often a case of the targeted use of carbon fibres. In particular, the superior strength, stiffness and resistance to environmental degradation of carbon fibres mean that they will find application in elements of the structure such as spar caps. In this capacity, they enable a more slender, higher performance blade than would otherwise be the case. In the quest to minimise the unit cost of power, a small increase in performance is worth a lot. In general, blades are assembled using a structural epoxy adhesive.

Both prepreg and resin infusion processes may be appropriate, depending on the detail of the requirement. Whilst the infusion process tends, in general, to result in a lower cost of material per kg and a simplified supply chain, the selection of prepreg materials would be expected to result in improved moulding consistency and a considerable reduction of development risk, particularly in cases where it is necessary to use carbon fibres. It is also the case that one or other of the processes may lend itself more naturally to the moulding of a given structural form. For example, if it is necessary to turn a mould upside-down during the production process, then a dry fabric preform intended for resin infusion will probably fall out of the tool, where a stack of sticky prepreg plies would be more likely to stay where they are needed-the reason is not always complicated!

In terms of configuration, AELs experience to date has predominantly been with MW-scale, variable pitch, hub-mounted blades. The favoured general configuration in this case (Figs. 2 and 3) is a central box-spar mounted in airfoil section skins/fairings. A rear web member provides additional reinforcement to the trailing edge. This arrangement confers a number of key advantages:-

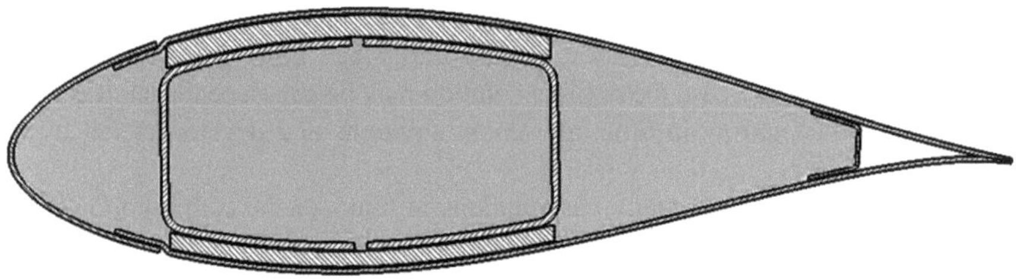

Fig. 2 MW scale, variable pitch blade section—earlier example for proof of concept machines

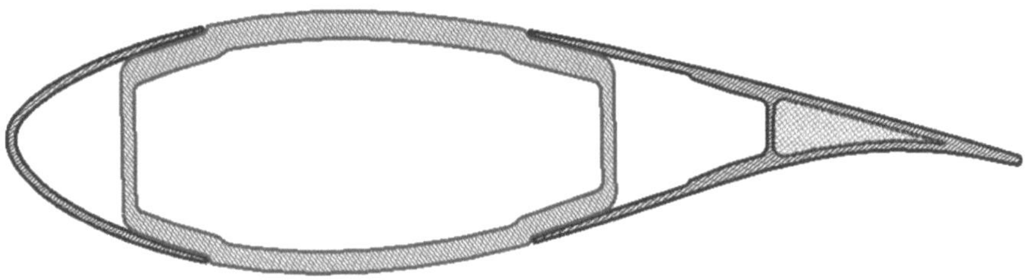

Fig. 3 MW scale, variable pitch blade section—evolved for production

- Neither the leading edge (vulnerable to impact) nor the trailing edge (relatively weak) is required to sustain serious shear loads.
- The primary structure (box-spar) is not exposed to unquantified direct leading edge impact.
- The box-spar can readily transition into a circular section at its root, where it will mate with the pitch bearing.
- In order to achieve a good hydrodynamic performance, it is important that the blade chord (distance from leading edge to trailing edge) is permitted to increase very rapidly as we move outboard of the blade root. It is straightforward to achieve this with the box-spar scheme, without requiring sudden changes of the section in the primary structure.

Whilst AEL's general scheme has tended to remain constant, there has been a considerable evolution over the years in terms of the detail, reflecting the changing requirements of developers.

In the early days, the priority would be to provide 2 or 3 blades relatively quickly for a 'proof of concept' prototype machine, whilst keeping tooling and development cost, and development risk to a minimum. Unit cost, part count and lead time to produce an individual blade were therefore of lesser importance than the overall project cost. A scheme appropriate to this earlier developmental stage is shown in Fig. 2. Whilst the relatively large number of mouldings and assembly stages means that such a scheme would not be well suited to cost-effective volume production, each moulding is in itself relatively straightforward to tool and manufacture. Therefore tooling costs are relatively low, and the chances of project costs ballooning due to development trouble are minimised.

The requirements of a blade designed for volume production are significantly different. By the time developers contemplate volume production, they should be more robustly funded and understand that it is necessary to absorb higher tooling and development costs, and some development risk in the quest for a low blade unit cost. Therefore it is required of the blade designer to be more ambitious in terms of the mouldings produced. Figure 3 shows substantially the same configuration as Fig. 2, but consists of 3 mouldings as opposed to 8. Although this development will assist greatly in the achievement of reduced blade unit cost and lead time, the mouldings are not as simple to engineer. In particular, the development of a successful single piece box-spar concept absorbed many hours of engineering time and the creation of several sub-scale prototypes.

Qualification of Materials and Structure

The general approach to the design and qualification of tidal energy blades calls for the design process to be synchronised with a 'pyramid' type test programme (Fig. 4) to determine or confirm relevant material and component properties and performance.

As such, the results of a large number of small, relatively cheap coupon tests are built upon by increasingly complex element, detail and full-scale structural tests. As the complexity of the test increases, so the number of tests diminishes.

In general, the lower levels of the pyramid (1 and 2) will tend to inform the final design, and the higher levels (3 and 4) serve primarily as confirmation that the final design and underlying data and assumptions are adequate.

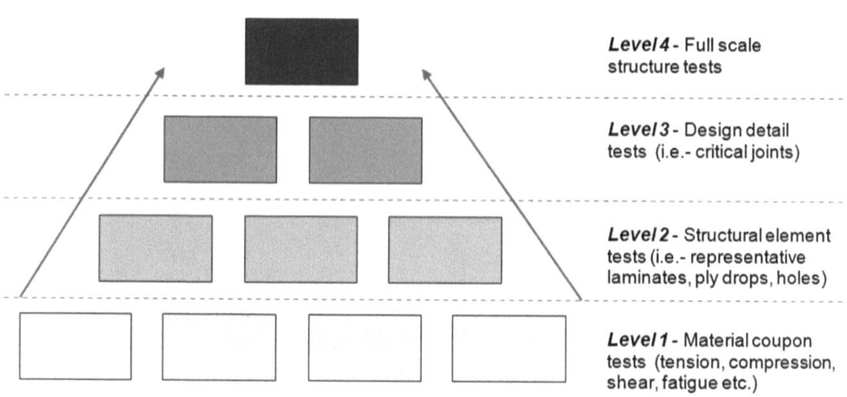

Fig. 4 Typical test pyramid

Perhaps unsurprisingly, the execution of a test programme is not without complication. Bearing in mind the size of tidal blades (8–9 m in length) and the heavy loading they endure (several MNm in terms of root bending moment), it may not be possible to locate a facility suitable for and willing to conduct a full-scale 'level 4' test. Wind blade test facilities are generally unsuitable, as they tend to employ winches for static loading, and resonant frequency excitation to induce fatigue loading.

In comparison to wind blades, tidal blades are relatively short and stiff, which means that it would be difficult to adequately control the applied load with winches. Furthermore, resonant frequency excitation is out of the question as a means of inducing fatigue, as the fundamental flap frequency of a tidal blade is much higher than the loading frequency at which mechanically induced heating would be expected in the structure. When it first became necessary to conduct a MW-scale blade test, AEL, therefore, found it necessary to build its own static test frame (Fig. 5). Load is distributed as appropriate along the length of the blade using a series of hydraulic rams and spreader beams, and the frame may be disassembled for storage. To date, static load tests have been successfully conducted at AEL on 2 separate blade designs.

But it is not just at the 'full-scale test' end of the test programme that complications are found. With regard to the effects of seawater saturation, it is generally far more practical to consider these at an early stage in the qualification process—typically as a part of the level 1 test programme. However, a seawater-ageing phase is usually very difficult to fit in the development programme for a new product (Fig. 6): One set of mechanical test coupons would normally be tested in the air with no ageing. Another set would be aged for several months at 45 °C in natural seawater, before mechanical testing in air. The design values would then be based on the least favourable data, adjusted as appropriate depending on the results of the level 2 programme. Coupons are designed to be as thin as possible (to minimise

Fig. 5 Tidal energy blade static test at AEL

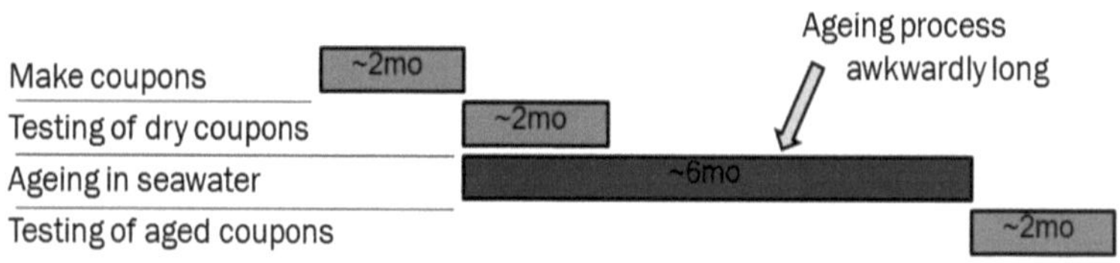

Fig. 6 Typical coupon testing timescales

time to saturate) but there are other considerations that limit achievable 'thinness' such as the avoidance of buckling in the case of compression tests, or the need to produce balanced and symmetrical laminates with what can be relatively thick plies.

The MARINET 'SEADEG' study was instigated as a collaborative effort between IFREMER, the University of Bristol and AEL to determine whether there exists any scope for further accelerating the seawater ageing of composite coupons. It follows that success in this aim could dramatically reduce the timescales required for the execution of the level 1 programme.

Another complication is the choice of a test environment for the level 1 tension fatigue coupons aged in seawater. The effect of the marine environment on the durability of composite materials has been studied in some detail since the work of Smith et al. [1]. Recent references [2–4] give an overview of current knowledge. Seawater ageing is known to result in potentially substantial reductions in the tensile strength [5] and tension fatigue performance of glass fibre composites tested in both air [6, 7] and seawater [2, 8]. Kosuri and Weitsman [9] showed that the tension fatigue performance of seawater-aged carbon composite laminates (AS4/3501-6 [0/90_3]s) may also depend on the test environment. They found that some laminates, when aged in seawater and then fatigued also in seawater demonstrated reduced performance when compared to those aged in seawater but then fatigued in the air. The proposed mechanism for this accelerated degradation (Fig. 7) is additional to the ageing affect that is normally quantified during materials qualification programmes as follows:

#1 -Transverse cracking of off-axis plies early in the fatigue life of a multiaxial laminate.
#2 -Water ingresses into transverse cracks as they open up under tensile loading
#3 -Pressurisation of the water during unloading, causing delamination between plies to grow more rapidly than would otherwise be the case.

However, the body of published experimental data relating specifically to this effect appears to be very limited. Therefore the study also aimed to identify whether such an effect would be apparent in a material/laminate typical of tidal energy blades.

The following sections describe the detail and outcome of a work programme undertaken to address some of the issues described above.

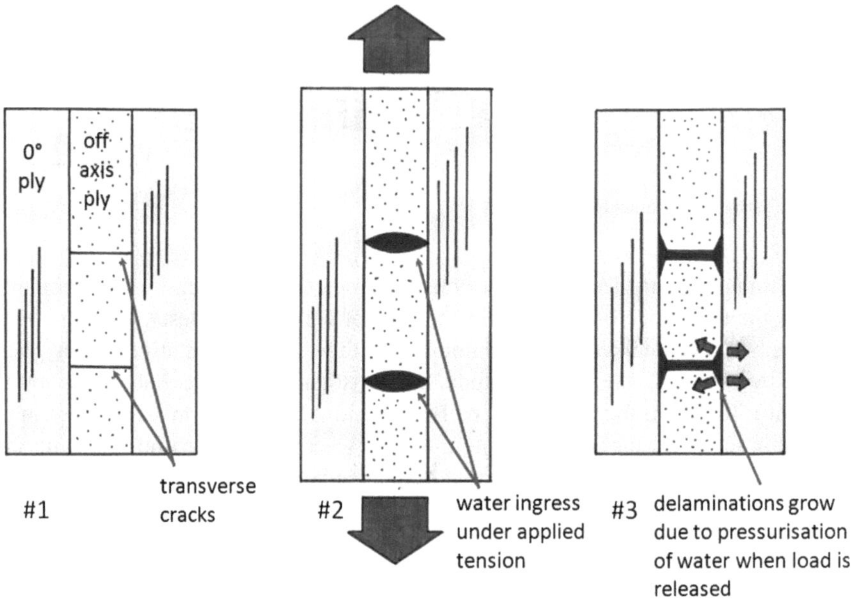

Fig. 7 Proposed mechanism by which fatigue testing in seawater may mechanically accelerate the growth of delaminations

2 Study Objectives

The objectives set for the body of work described in this chapter were, therefore, as follows:

- To determine whether the choice of conditioning medium (natural seawater or deionised water) would affect the moisture absorption characteristics, fibre direction tensile strength (σ_{11T}) and interlaminar shear strength (τ_{13}) of infused and prepreg unidirectional glass/epoxy laminates. These two parameters were selected as being good indicators, respectively, of fibre and matrix/interface performance and degradation.
- To quantify the effect of pressure and temperature on the rate at which the infused and prepreg glass/epoxy laminates absorb moisture.
- To determine whether the choice of the test environment (natural seawater or air) would have a noticeable effect on the tension fatigue strength of [0/+ 45/0/-45]s infused glass/epoxy laminates already aged in natural seawater until close to saturation.

3 Procedures

Materials for the programme were chosen to be typical of those used in the construction of tidal blades.

Resins: As discussed in Sect. 1, the manufacture of tidal blades may require the use of prepreg and/or infusion resins. Therefore the majority of this study was conducted using representative examples of both.

Fibres/fabric: As discussed in Sect. 1, glass fibre would normally be selected for tidal blade construction in preference to carbon fibre, unless the latter was specifically required for its superior strength, stiffness or environmental resistance. Therefore glass fibres were deemed most appropriate, particularly as their greater sensitivity to environmental effects would make for clearer results. More specifically, the selected fabric comprised 'Advantex' fibres in a 567 g/m^2 unidirectional non-crimp format.

Test coupons were manufactured at AEL with an 80 °C ultimate cure temperature, then shipped to IFREMER for conditioning and mechanical testing (Table 1).

In order to establish control data, Quasi-static mechanical tests (σ_{11T} & τ_{13}) and tension fatigue tests (σ_{XXT}) were conducted on dry (un-aged) coupons according to Table 2, (M1, M4a).

Table 1 Test matrix—coupon conditioning

			Materials				
ID	Subject	Medium	Infused	Prepreg	Temperature (°C)	Pressure (bar)	No. coupons/test
Cla	Effect of conditioning temperature on laminate weight gain in natural seawater	Natural seawater	Yes	Yes	4	1	3
Clb			Yes	Yes	25		
Clc			Yes	Yes	40		
Cld			Yes	Yes	60		
Cle			Yes	Yes	80		
C2a	Effect of conditioning medium on laminate weight gain	Deionised water	Yes	Yes	40	1	3
C3a	Effect of conditioning pressure on laminate weight gain	Tap water	Yes	Yes	40	1	3
C3b			Yes	Yes		500	5
C4	Weight gain check for thicker |0/ + 45/0/-45]s fatigue coupons	Natural seawater	Yes		40	1	1

Table 2 Test matrix—mechanical testing

			Materials				
ID	Subject	Test type	Infused	Prepreg	Conditioning environment	Test environment	No. coupons/test
M1a	Mechanical properties, dry (unconditioned) control	σ_{11T}	Yes	Yes		Air	5
M1b		τ_{13}					5
M2a	Mechanical properties, after conditioning in seawater	σ_{11T}	Yes	Yes	Natural seawater, 8 months @40 °C	Air	5
M2b		τ_{13}					5
M3a	Mechanical properties, after conditioning in deionised water	σ_{11T}	Yes	Yes	Deionised water 8 months @40 °C	Air	5
M3b		τ_{13}					5
M4a	Laminate tension fatigue, effect of test environment on conditioned [0/+ 45/0/-45]s laminates	Tension fatigue R = 0.1 1 Hz	Yes			Air	6
M4b			Yes		Natural seawater, 8 months @ 40 °C	Air	6
M4C			Yes		Natural seawater, 8 months @ 40 °C	Natural seawater	10

Coupon conditioning and regular weighing were carried out in accordance with the procedures explained in Sect. 3.2. Weight gain coupons were aged according to the conditioning matrix in Table 1 to establish the effects of conditioning temperature (C1), conditioning medium (C2) and conditioning pressure (C3) on their moisture absorption characteristics.

Quasi-static mechanical test coupons (σ_{11T} and τ_{13}) were conditioned in natural seawater and deionised water at 40 °C for a period of almost 8 months. Tension fatigue coupons (σ_{XXT}) were conditioned in natural seawater at 40 °C for 9 months. Mechanical tests were then conducted on these conditioned coupons according to the test matrix in Table 2.

3.1 *Design and Manufacture of Coupons*

Test panels were manufactured using standard AEL methods, developed to be as representative as possible of full-scale manufacturing processes. Prepreg panels were manufactured by oven-curing the prepreg under the vacuum between a flat tool and a caul plate. Infused panels were manufactured by vacuum-infusion of the fabric stack between a flat tool and a caul plate. The panels were allowed to cure at room temperature for approximately 24 h before transfer to an oven for final cure.

Subsequently, end tab strips were bonded to the tension (σ_{11T}) and fatigue (σ_{xxT}) coupons using epoxy adhesive film and a 70 °C cure cycle. A water-cooled diamond wheel saw was used to cut coupons from the panel.

Materials specifications for both the infused and prepreg coupons are shown in Table 3. Coupon laminate details and dimensions are presented in Table 4.

Table 3 Material specifications

Matrix:-	Infused	Prepreg
Reinforcement	567gsm Advantex SE2020 Unidirectional non-crimp (587gsm including carrier fibres and stitching)	
Initial cure	24 h @ room temp	N/A
Final cure	6 h @ 80 °C	6 h @ 80 °C
Tg onset (dry)	~70 °C (DSC)[a]	89 °C (DMA)[b]
Tg peak tan delta	~89 °C (DMA)[a]	101 °C (DMA)[b]
Typical cured ply thickness	0.442 mm	0.488 mm
Typical Fibre volume fraction (structural)	49.0%	45.6%
Typical Fibre volume fraction (total)	50.7%	47.2%

[a]Data supplied by manufacturer for cure cycle with 24 h @ room temp, 16 h @ 50 °C, 2 h @ 80 °C
[b]Data supplied by manufacturer for identical cure cycle

Table 4 Coupon details

Coupon type:-	Tension σ_{11T} (M1a-M3a)	Interlaminar shear—τ_{13} (Mlb-M3b)	Fatigue σ_{xxT} (M4)	Weight gain (standard) (C1-C3)	Weight gain (thick) (C4)
Layup	0_4	0_5	[0/+ 45/0/-45]s	0_4	[0/+ 45/0/-45]s
Nominal width (mm)	25	10	25	50	50
Overall length (mm)	240	16	240	50	50
Gauge length (mm)	120	12	120	N/A	N/A

3.2 Conditioning and Weighing of Coupons

Coupons (C1, C2, C4) were aged in either natural seawater or deionised water as appropriate (Table 1). Tank capacity in each case was 170 litres. The temperature was monitored and recorded, and the water was continuously renewed. The coupons were weighed periodically on a Sartorius balance: After removal from the tanks, excess surface water was wiped off with absorbent paper and the coupons were weighed before being replaced in the tanks. The relevant mechanical test coupons (Table 2, M2–M4) were aged in the same tanks, then removed at the end of the ageing period and held at 20 °C in seawater or deionised water (corresponding to the conditioning medium) until testing. To enable ageing under hydrostatic pressure (C3b), coupons were placed in tap water in pressure vessels inside temperature controlled ovens. Control coupons (C3a) were placed in a tap water bath at the same temperature, in the oven next to the pressure vessel. All measurements were made using equipment calibrated annually by a certified external calibration company.

3.3 Mechanical Testing

Tensile tests were performed on a 200 kN electro-mechanical test frame according to ISO 527. Interlaminar shear tests (short beam shear) were executed in accordance with ASTM 2344. In both cases, tests were performed under displacement control at 2 mm/minute.

Tension fatigue tests were executed on a 250kN servo-hydraulic test frame. The chosen loading frequency was 1 Hz, after preliminary measurements made with an infrared camera, (Fig. 8a), had indicated heating effects at higher frequencies in air. Tests were performed under load control at R = 0.1 (cyclic minimum load/cyclic

Fig. 8 Tensile fatigue test setup, **a** in air, with IR camera, **b** in natural seawater

maximum load = 0.1). A special fixture was built, (Fig. 8b) in order to perform tests in seawater. This allowed temperature controlled natural seawater (25 °C) to be circulated around the specimen during the test.

4 Results

4.1 Variation of Natural Seawater Temperature

Immersion tests of up to 2 years in duration confirmed the expectation that an increase in conditioning temperature greatly accelerates the rate of coupon weight gain (Figs. 9a, b). The infused coupons were seen to saturate at 0.9–1% weight gain. However, in the case of the prepreg coupons, only the 60 °C curve seems to have stabilised, appearing to reach saturation at 1.4% weight gain.

At all temperatures, the infused coupons were seen to reach saturation much more rapidly than the prepreg. For both materials when aged at 80 °C, colour changes and further changes in weight after apparent saturation were noted (weight gain for the infused coupons, weight loss for the prepreg). This suggests that some manner of material degradation beyond the physical effects of moisture ingress may have taken place, (possibly oxidation) and appears to confirm the normal assumption that coupons should not be aged very close to or exceeding the matrix Tg, as this would risk altering the material in a manner unrepresentative of normal service.

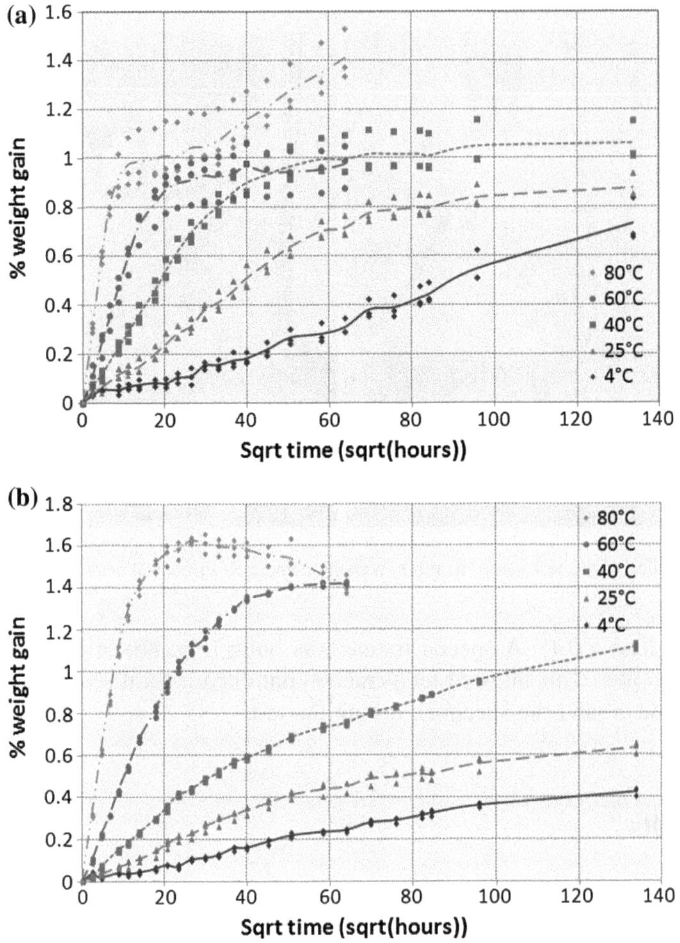

Fig. 9 **a**—Weight gain of infused coupons in natural seawater—effect of conditioning temperature. **b**—Weight gain of prepreg coupons in natural seawater—effect of conditioning temperature

It is possible for both materials to normalise the time required to achieve a certain weight gain at any temperature against the time required to achieve the same weight gain at 40 °C (Figs. 10a, b). Linear relationships can be seen to exist between the conditioning temperature and log[normalised time], particularly convincingly in the case of the infused coupons.

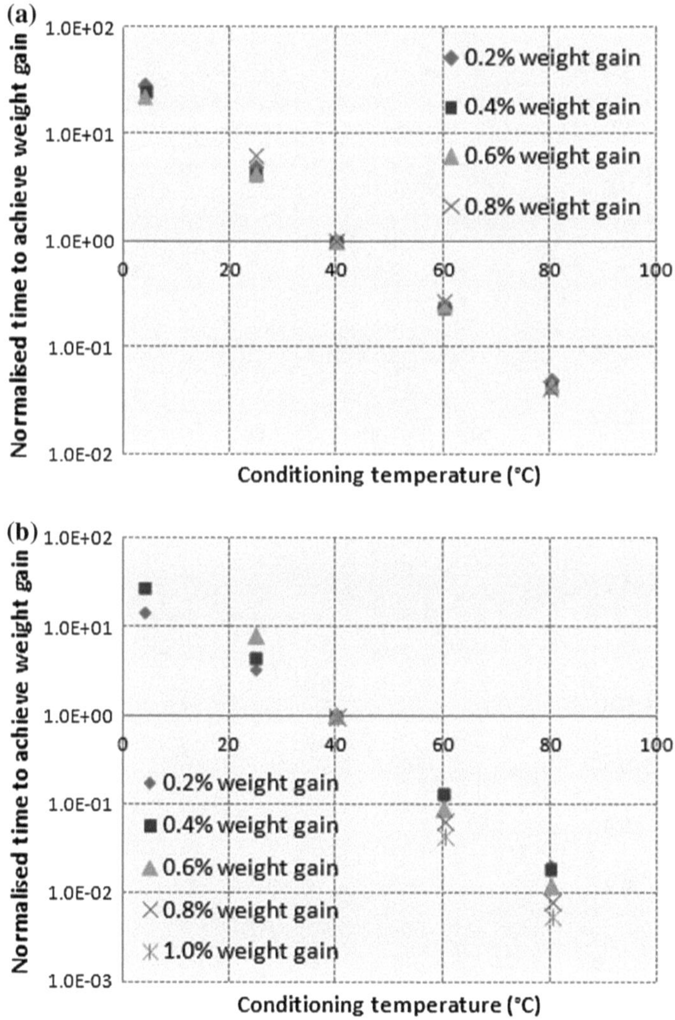

Fig. 10 **a**—Time to weight gain normalised against 40 °C data—infused coupons. **b**—Time to weight gain normalised against 40 °C data—prepreg coupons

4.2 *Variation of Pressure and Conditioning Medium*

The effect of increasing the ambient pressure from 1 to 500 bars is very limited for both types of the coupon (Figs. 11a, b). For the infused coupons, the effect of high pressure even appears to be to retard the initial rate of weight gain. It is suggested in other work where strong pressure effects on diffusion kinetics are noted [10], and that these effects are related to relatively high-laminate porosity. Considering the logistics of a normal accelerated ageing programme, it is not too great a

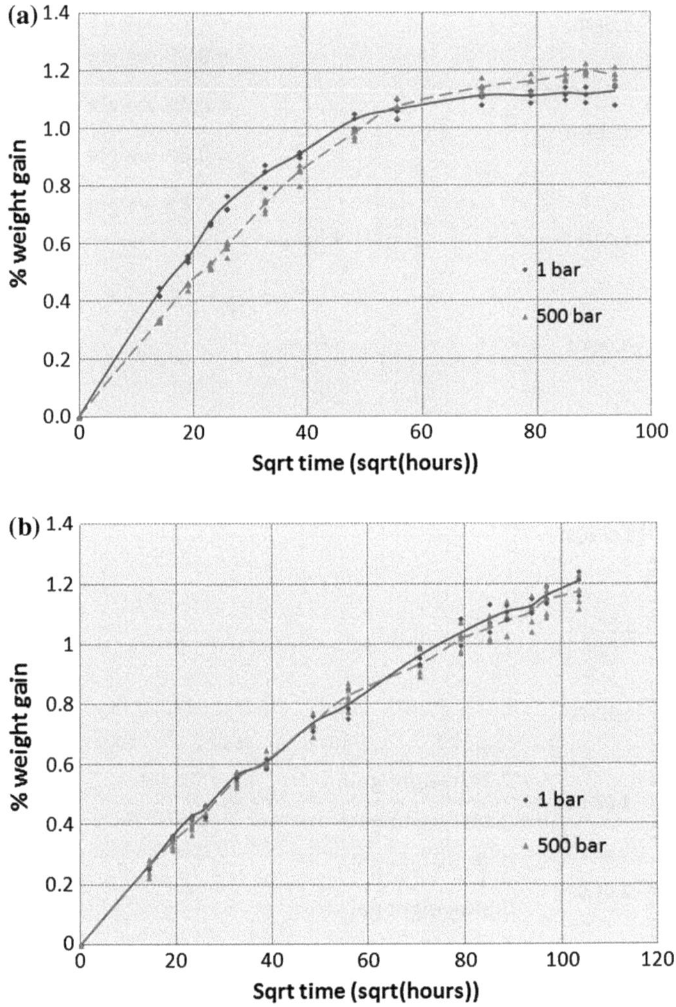

Fig. 11 **a**—Effect of pressure—infused coupons in tap water at 40 °C. **b**—Effect of pressure—prepreg coupons in tap water at 40 °C

disappointment that the use of hyperbaric equipment has very little effect. The practicalities of depressurising and re-pressurising the conditioning unit every time access was required, and the likely capital and maintenance costs of the equipment would be particularly onerous.

With regard to the effect of conditioning medium, initial rates of weight gain are similar. However, it can be seen that coupons conditioned in deionised water will ultimately achieve a greater weight gain than those conditioned in seawater (Figs. 12a, b)

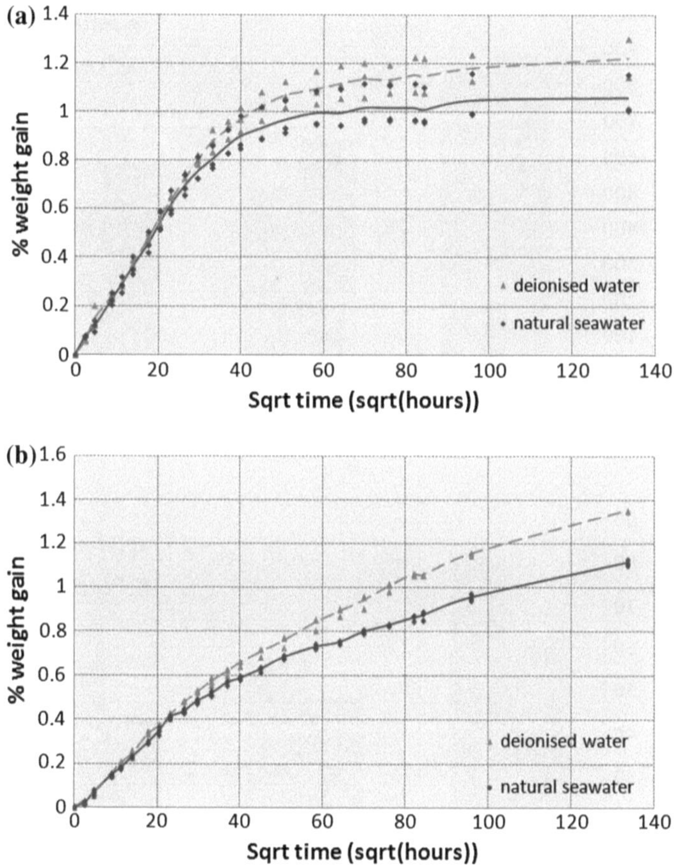

Fig. 12 **a**—Effect of conditioning medium on weight gain—Infused coupons at 40 °C. **b**—Effect of conditioning medium on weight gain—Prepreg coupons at 40 °C

4.3 *Effect of Conditioning Medium on Mechanical Properties*

Both conditioning media clearly have an adverse effect on the tensile and interlaminar shear strengths of the infused and prepreg coupons (Figs. 13 and 14) with the deionised water resulting consistently in greater degradation. This means that changing the conditioning medium from seawater to deionised water would risk building too great a seawater knockdown factor into the design allowables, possibly resulting in an uncompetitive product. Note that the higher tensile strength of the infused samples is due to the higher fibre content (see Table 3).

As all coupons were manufactured using the same fabric, it is worth considering (Figs. 15 and 16) whether there could be a correlation between weight gain and strength when normalised with respect to the dry values. If it were possible to

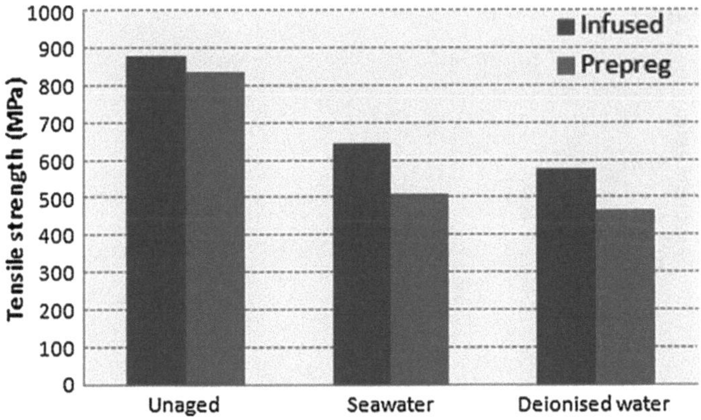

Fig. 13 Fibre direction tensile strength—effect of 8 months ageing @ 40 °C

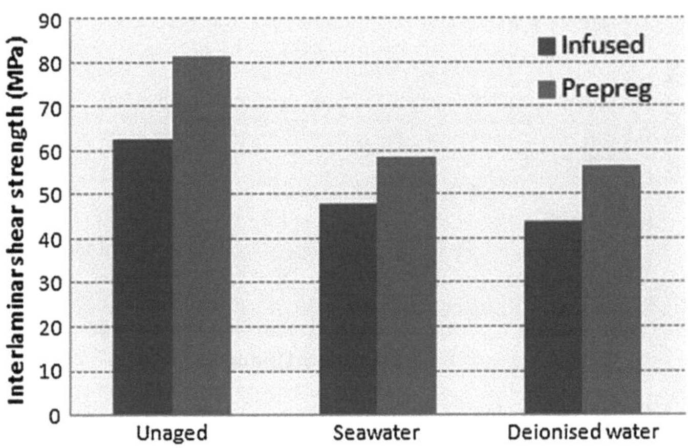

Fig. 14 Interlaminar shear strength—effect of 8 months ageing @ 40 °C

identify a relationship independent of conditioning medium and resin option, it could prove a very valuable aid to resin selection. A plot of the normalised tensile strength (σ_{11T}) against weight gain at test (estimated from the weight gain coupon data), does not show any correlation. However, if instead we plot σ_{11T} against weight gain at saturation, then it is possible to observe a reasonable linear correlation (Fig. 15) between these two parameters. Of course, the material close to the surface will be saturated long before the coupon as a whole, so the observation does make sense. However, we do not see such a convincing correlation when we plot the interlaminar shear strength (τ_{ILS}) against weight gain (Fig. 16), either at test or saturation, and it should be noted that some authors have proposed more complex correlations in the past [11].

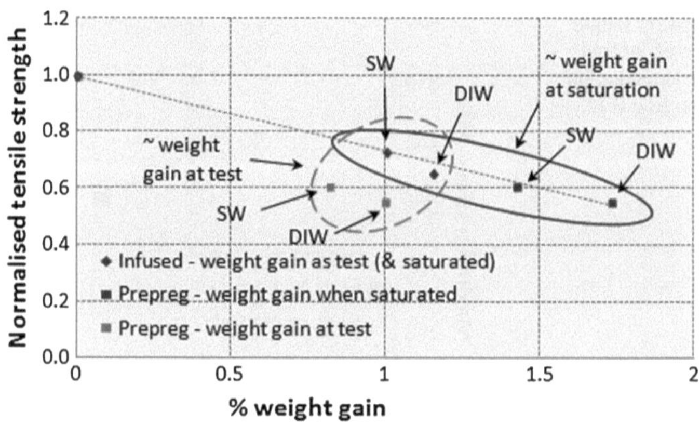

Fig. 15 Normalised tensile strength against weight gain

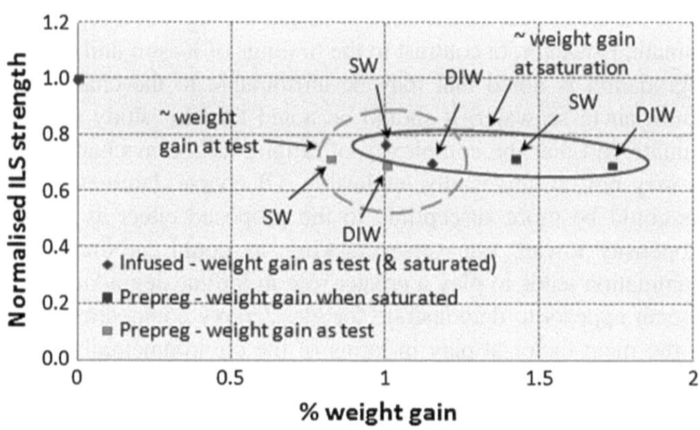

Fig. 16 Normalised ILS strength against weight gain

(Note that the weight gain at saturation for the prepreg coupons in deionised water was estimated by factoring weight gain at saturation in seawater by the ratio of *weight gain in deionised water: weight gain in seawater* at the test.).

4.4 *Effect of Ageing and Test Environment on Tension Fatigue Strength of a [0/+ 45/0/-45]s Glass/Epoxy Laminate*

When tested after 9 months' immersion in natural seawater at 40 °C, results show that the seawater-aged tension fatigue coupons would have been saturated or very close to saturation (1% weight gain). Figure 17 confirms that seawater ageing also

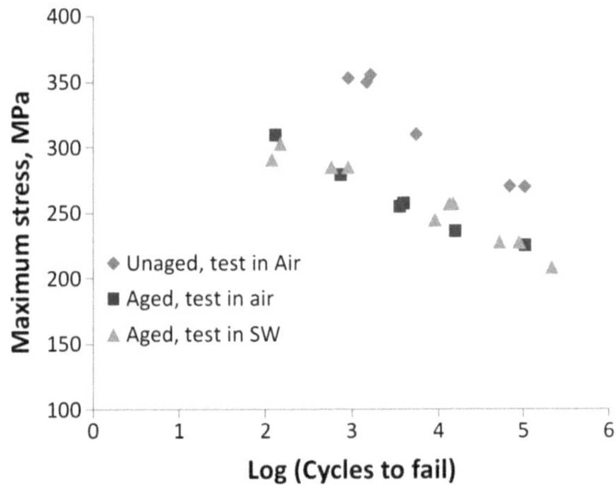

Fig. 17 Tension fatigue performance of infused glass/epoxy laminate, [0/ + 45/0/-45]s, R = 0.1

has a notable adverse effect on the tension fatigue strength of a representative glass epoxy laminate. However, in contrast to the findings of Kosuri and Weitsman [9] no further degradation is noted that may be attributable to the change in test environment from air to seawater. It should be noted that this study only applies to a single laminate and that the complexity of fatigue failure mechanisms means that this result may not be universally applicable. Of course, laminates including 90° cross plies could be more susceptible to the proposed effect as a result of their greater propensity toward transverse cracking, as could carbon/epoxy laminates where delamination tends to play a greater role in fatigue degradation. In any case, the experiment appears to demonstrate for glass/epoxy composites that the ageing process is the main factor at play in terms of the environmentally induced degradation of tension fatigue performance.

5 Conclusions

Bearing in mind the concerns discussed in Sect. 1 with regard to the seawater ageing and testing of laminates for tidal blade construction, it is possible to draw some very useful conclusions from this study.

- Ageing at high pressure should be ruled out as a means of accelerating the ageing process. It does not meaningfully change the rate of ingress of moisture, nor is it a particularly practical proposition.
- Ageing in deionised water may slightly reduce the time required to achieve a given level of saturation, but it would also appear to have a greater adverse effect on mechanical performance than ageing in natural seawater. Therefore, natural seawater remains the conditioning medium of choice.

- The only practical option (of those examined) to further accelerate ageing would be an increase in conditioning temperature. However, the benefits of such a change could be significant; by increasing the standard conditioning temperature towards 60 °C, seawater-ageing timescales could be reduced by a factor of three or more. This is consistent with a 4 month saving in the duration of a qualification programme. Of course, this would depend on the specific resin system and Tg; care would need to be taken in setting a temperature limit below which any degradation effects would be representative of those anticipated in normal service.
- There can be significant differences between resins in terms of weight gain at saturation, and time to saturate.
- There appears to be a linear correlation between weight gain at saturation and fibre direction tensile strength normalised with respect to the dry value, for coupons with identical Advantex glass reinforcement, but variations in matrix and ageing medium,
- For a representative [0/+45/0/-45]s Advantex glass laminate, there is a significant knockdown in terms of tension fatigue performance associated with ageing in seawater However, there is no further knockdown associated with conducting the fatigue test in seawater, as opposed to air.

6 Further Work

Before any change can be implemented in terms of conditioning temperature, it will be necessary to verify that an increase in conditioning temperature (as required to accelerate the ageing process) does not confer any unexpected effect on mechanical performance. Furthermore, it would be interesting to follow up the observation of linear dependency of normalised tensile strength on weight gain at saturation. There could be significant beneficial implications for materials screening or qualification programmes, should this relationship be shown to hold more widely.

It would also be worthwhile to extend the fatigue study to consider a greater variety of laminates and materials, as it is possible that other aged laminates (particularly those including 90° cross plies and/or carbon reinforced laminates) could be more susceptible to accelerated fatigue damage when the fatigue test is conducted in seawater.

Acknowledgements The authors wish to acknowledge the support afforded by the European funded MARINET (Marine Renewables Infrastructure Network) programme, which has enabled this study and promoted valuable cross-border collaboration.

References

1. Smith CS (1990) Design of marine structures in composite materials, Elsevier
2. Weitsman Y (2012) Fluid effects in polymers and polymeric composites. Springer, New York, Dordrecht, Heidelberg, London
3. Davies P, Rajapakse Y (ed) (2013) Durability of composites in a marine environment, Springer
4. Harper P, Hallett SR, Fleming A, Dawson M (2016) Advanced fibre-reinforced composites for marine renewable energy devices. In: Chapter in marine applications of advanced fibre-reinforced composites. Woodhead Publishing
5. Boisseau A, Davies P, Thiebaud F (2012) Sea water ageing of composites for ocean energy conversion systems: Influence of glass fibre type on static behaviour. Appl Composi Mater 19 (3–4): 459–473
6. Dawson M (2011) AEL internal report number MTR097A
7. Sun L (2014) AEL internal report number MTR207B
8. Boisseau A, Davies P, Thiebaud F (2013) Fatigue behaviour of glass fibre reinforced composites for ocean energy conversion systems. Appl Compos Mater 20(2):145–155
9. Kosuri R, Weitsman Y (1995) Sorption process and immersed fatigue response of GR/EP composites in seawater. In: Proceedings of ICCM-10, Whistler, B.C., Canada
10. Humeau C, Davies P, Jacquemin F (2016) Moisture diffusion under hydrostatic pressure in composites. Mater Des 96:90–98
11. Pritchard G, Speake SD (1987) The use of water absorption kinetic data to predict laminate property changes. Composites 18(3):227–232

Influence of Composite Fatigue Properties on Marine Tidal Turbine Blade Design

Vesna Jaksic, Ciaran R. Kennedy, David M. Grogan, Sean B. Leen and Conchúr M.Ó. Brádaigh

Abstract The structural design of marine tidal turbine blades is governed by the hydrodynamic shape of the aerofoil, extreme loadings and composite material mechanical properties. The design of the aerofoil, chord and twist distribution along the blade is generated to optimise turbine performance over its life time. Structural design gives the optimal layout of composite laminae such that ultimate strength and buckling resistance requirements are satisfied. Most structural design approaches consider only extreme static loads, with a lack of dynamic load-based fatigue design for tidal blades. Approaches for tidal turbine blade design based on dry and immersed composite material fatigue life are studied.

Keywords Composites · Marine environment · Fatigue properties · Tidal turbine blades · Design

V. Jaksic (✉)
Sustainable Infrastructure Research & Innovation Group (SIRIG), Civil, Structural and Environmental Department, Cork Institute of Technology, Cork, Ireland
e-mail: vesna.jaksic@cit.ie

C.R. Kennedy · D.M. Grogan · S.B. Leen
Department of Mechanical Engineering, National University of Ireland, Galway, Ireland
e-mail: ciaran.kennedy@nuigalway.ie

S.B. Leen
e-mail: sean.leen@nuigalway.ie

C.M.Ó. Brádaigh
Institute for Materials and Processes, University of Edinburgh, Edinburgh, Scotland, UK
e-mail: c.obradaigh@ed.ac.uk

P. Davies and Y.D.S. Rajapakse (eds.), *Durability of Composites in a Marine Environment 2*, Solid Mechanics and Its Applications 245,
https://doi.org/10.1007/978-3-319-65145-3_11

1 Introduction

In order to fully utilise the tidal blade structural material fatigue life it is necessary to have information on its performance in marine environment for the full design life of marine renewable energy devices (10–20 years). To date, however, there are no reports available of experience with heavily loaded GFRP structures operating in this environment. In addition, the literature on fatigue test programmes for the composite materials that will be used in marine renewable energy devices is limited. Consequently, designers of these devices are forced to be quite conservative in their structural design, leading to increased cost. A comprehensive fatigue life model for composite blades, incorporating realistic hydrodynamic loadings, cyclically varying blade stresses and wet composite material fatigue properties, will therefore be very valuable for tidal turbine designers. In this regard, the aim of the study is to quantify the degradation in fatigue strength of composite materials due to water uptake and to predict the degradation in tidal turbine blade life due to water saturation.

1.1 Tidal Energy

Tidal energy is gaining increased attention due to its predictable nature as renewable energy source. The efficiency of tidal power is around 80%, much greater than coal, solar or wind. The geographic distribution of tidal energy resources around the world is presented in Fig. 1 [1].

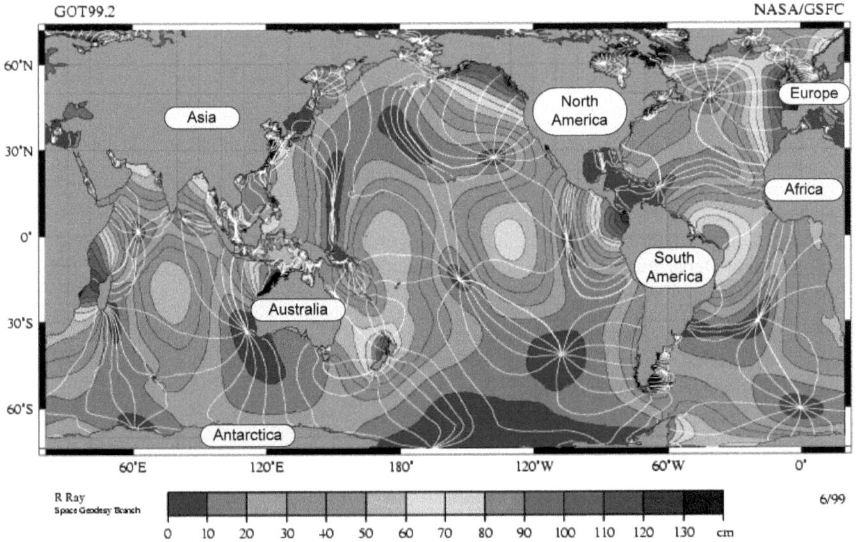

Fig. 1 World tidal energy patterns [1]

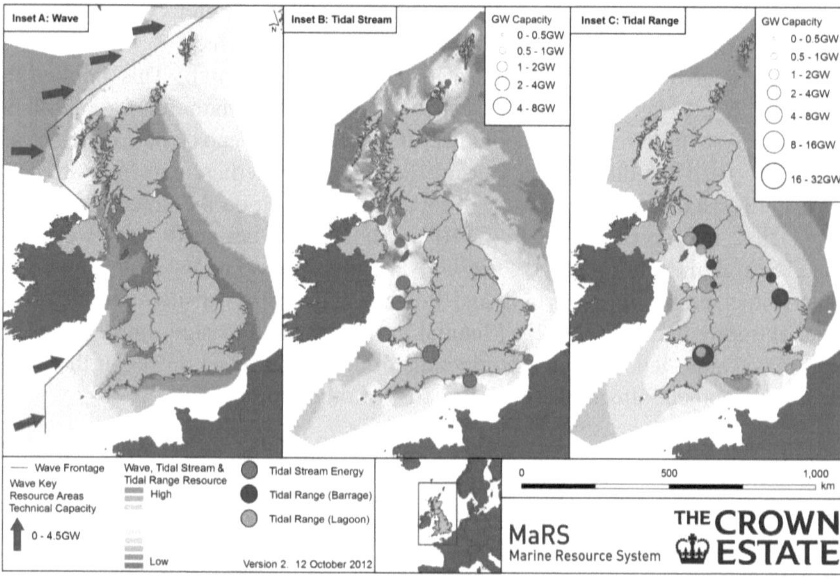

Fig. 2 Wave and tidal power distribution and energy map—UK [2]

The map shows that there is considerable tidal energy-harnessing potential around the coastlines of the UK, Ireland and France, among others. The total theoretical tidal energy resources available in broad geographic areas around the UK (Fig. 2) are estimated to be 216 TWh/year [2] while estimated Irish resources are 230 TWh/year [3]. However, technological limitations, physical, environmental and commercial constraints will make extraction of all this available energy impossible. However, taking into account the ocean energy projects which have been awarded funds, Europe could see up to about 57 MW of tidal energy capacity installed operational by 2020 [4].

1.2 Composite Tidal Turbine Blades

Despite the predictable nature of tides, tidal current intensity and variability over small geographical areas cannot be foreseen [5]. Turbine rotors are the part of tidal device structures that are most sensitive to harsh marine environment, extreme weather and turbulence flows [6, 7], as well as having to withstand erosion due to ice, sand and floating trees. Tidal turbine blades are subject to water ingress and saturation during the device employment period [8]. The blade failures on a number of prototypes highlight the need for design that will withstand the expected hydrodynamic loads [9, 10]. Experience shows that blade failure due to environmental loading often happens in short periods. Structural health monitoring of tidal blades on seabed-mounted devices, in particular, is difficult and expensive and in the case of damage, costly lifting vessels need to be employed.

Hence, tidal blades require high static and fatigue strength. This implies the necessity to use thick composite sections or higher strength reinforcing fibres (carbon). Glass fibre-reinforced polymers (GFRP) such as epoxy or vinyl ester are commonly used for tidal blades due to their good properties, e.g. high strength, corrosion resistance and low cost. However, the research on the potential of using carbon fibre-reinforced polymers (CFRP) in tidal blade design [9, 10] is ongoing [11]. The polymers normally used in GFRP can absorb up to 5% water by weight when immersed for long periods and this can reduce the tensile strength of the material more than 25% [12]. Material properties and design impact the life expectancy, and hence energy production from tidal turbines [13]. Furthermore, blade development affects the load and cost of other components.

1.3 Modelling of GFRP Fatigue Life

Composites are a natural choice for turbine blades but there is little test data available on material behaviour under coupled environmental and cyclic loading. An extensive review of modelling of fatigue in GFRP has divided the work into three broad approaches [14]. First is a testing approach where life predictions are based on test data of the exact or similar material; second is the phenomenological approach where predictions are based on the stiffness and residual strength behaviour; and third is a progressive damage approach where damage in the unidirectional (UD) lamina is predicted and incremented until a final failure state is reached, thereby predicting fatigue life. The testing approach to fatigue life estimation is by far the most widely used [15, 16]. The technique is constantly being refined to include effects like spectrum loading and complex constant life diagram (CLD) results [17]. Stiffness of GFRP laminates degrades by between 10 and 20% during fatigue cycling. The main drawback of the models discussed up to this point is their lack of flexibility in dealing with different laminate layups and/or loading patterns. Micromechanical approaches that predict the response of the laminate based on damage mechanisms in the individual UD plies offer a potential solution. The simplest approach is to degrade the matrix properties based on observed levels of cracking [18] and use classical laminate theory (CLT) to integrate the results. Others have considered two damage mechanisms, namely matrix cracking and interlaminar delamination [19].

2 Blade Hydrodynamic and Fatigue Model

2.1 Tidal Turbine Blade Design Assumptions

The preliminary design of the tidal blade assumes that device operates in a steady flow. The device configuration considered for the modelling and benchmarking is based on a wind turbine, i.e. a machine which has a horizontal axis oriented up-tide and has three pitch-regulated (PR) blades [20]. Hydrodynamic design of tidal turbine blades can be expected to follow a similar pattern to that of wind turbine blades [9, 21]. Hence, the Blade Element Momentum (BEM) theory is employed for a blade hydrodynamic modelling [22, 23]. In this regard, a preliminary design methodology is presented, based on a minimum number of input parameters, in order to predict the fatigue life of the blades [24, 25].

2.2 Turbine Blade Fatigue Design Methodology

The preliminary fatigue design methodology for GFRP tidal turbine blades consists of a number of separate models, combined to facilitate the hydrodynamic and structural calculations required (Fig. 3).

The first module in the design process algorithm is a tidal model which predicts the tidal current speed at any time for specified local tidal velocities [9, 24]. The output from this tidal model forms a key input to a hydrodynamic model, which in turn computes the radial distributions of local relative blade-fluid velocities, axial and tangential blade forces, optimum chord length, and pitch angle. The subsequent structural module, based on development of a finite element model of the blade, and driven by the output from the hydrodynamic model, is then employed to

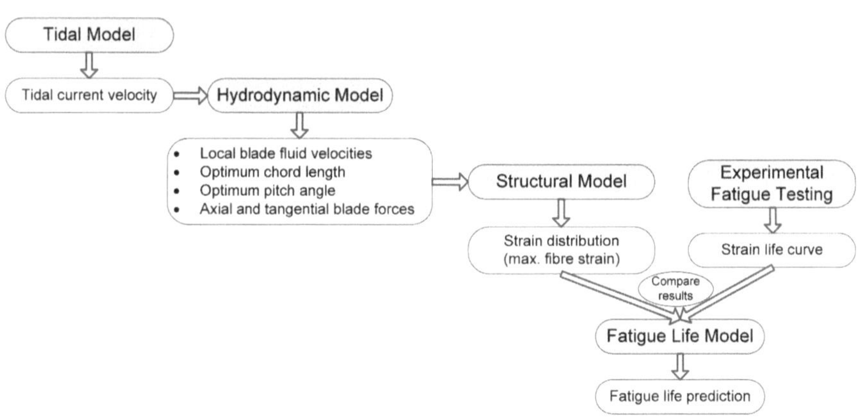

Fig. 3 Algorithm of tidal turbine blade design methodology

characterise the strain distribution in the turbine blade. The fatigue model accounts for each rotation of the blade explicitly and determines the maximum strain in the blade for each cycle. It then takes that maximum strain, compares them to experimentally determined strain–life curves for immersed and dry materials and obtains a damage fraction for that cycle [13].

The method can be applied to the preliminary investigation of a number of design aspects for tidal turbines. All of the models used in the methodology are relatively simple and can be quickly processed. This allows adjustment of the key parameters in order to study the effect of changes in the device design and performance. The model takes into the account the experimental results of fatigue testing on GFRP laminates, manufactured using materials and processes which are the current state of the art used in marine renewable energy devices.

2.3 Tidal Model

The tides are the result of interaction of the gravity of the sun, earth and moon. The fall and rise of tides create kinetic energy. The first module in the design process is a tidal model, which predicts the tidal current speed at any time for specified measured local tidal velocities (Fig. 4). The water speed depends on the local topography, but if the spring (highest tides) and neap (lowest tides) maximum velocities are measured, the full cycle can be approximated by a combination of a semi-diurnal sinusoid and a fortnightly sinusoidal function [26, 27]. The output from the tidal model forms a key input to a hydrodynamic model.

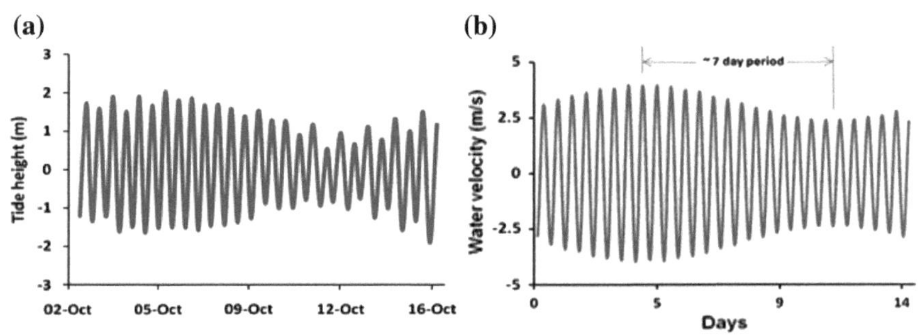

Fig. 4 **a** Typical tide height pattern around Ireland [28] and **b** Sinusoidal model output for water velocity

2.4 Hydrodynamic Model

The hydrodynamic model (HDM) generates an optimised blade shape for any size turbine when given water velocity (generated by tidal model) and number of blades. It also calculates the forces that are exerted on the single blade by the water flow. The HDM is based on a stream tube or blade element momentum approach [29, 30] (Fig. 5). Hence, for each stream tube (radial distribution of the blade), local relative blade-fluid velocities, axial and tangential blade forces, optimum chord length, and pitch angle can be calculated. Furthermore, the model predicts the maximum power output, torque and bending moment at the blade root at the chosen water velocity.

The lift and drag forces can be estimated to a first-order approximation using a stream tube momentum model. The model (Fig. 5) assumes a series of isolated concentric tubes within which momentum loss is in balance with the forces acting on the blade. The fluid forces on the blades (for steady-state operation), inside any stream tube, must equal the momentum lost from that stream tube.

The lift force (*L*), perpendicular and drag force (*D*), parallel to the fluid velocity, generated by a stream tube on the blade section inside it [24], can be calculated:

$$L = \frac{1}{2} C_L V_r^2 \rho S \Delta R \tag{1}$$

$$D = \frac{1}{2} C_D V_r^2 \rho S \Delta R, \tag{2}$$

where C_L and C_D are the coefficients of lift and drag for the aerofoil selected [31], V_r is apparent or relative velocity seen by the blade (Fig. 5), ρ is the density of the fluid, S is the chord length of the blade at the stream tube radius and ΔR is the radial thickness of the stream tube. The axial (thrust) force, F_A, and the tangential (torque) force, F_C, acting on the blade by observed stream tube can be obtained:

$$F_A = L \cos\theta + D \sin\theta \tag{3}$$

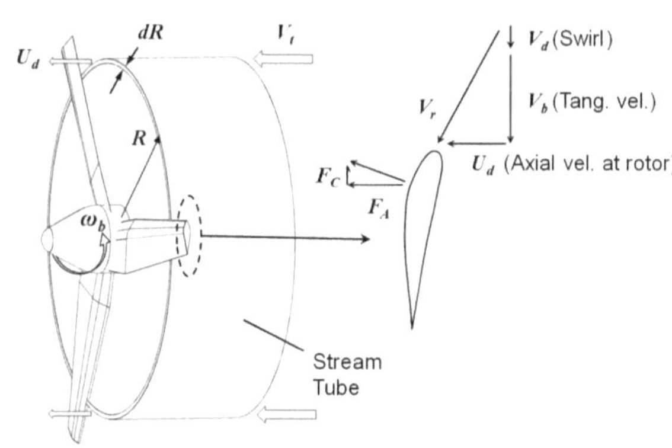

Fig. 5 Schematic of hydrodynamic model concept showing one of the *n* concentrically stacked stream tube annuli and the associated fluid flows [24]

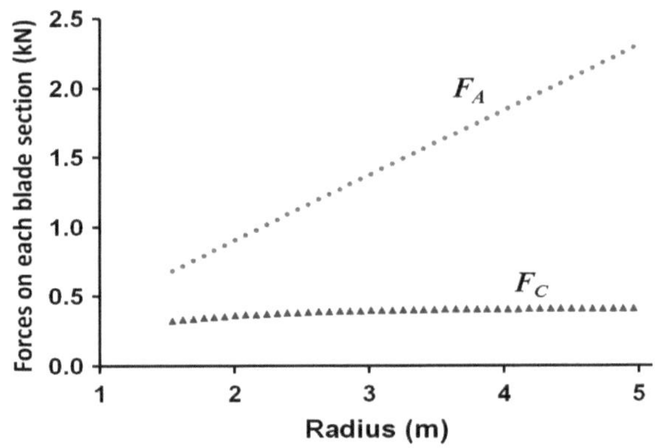

Fig. 6 Radial distributions of axial (F_A) and tangential (F_C) forces from hydrodynamic model at 2.6 m/s water velocity

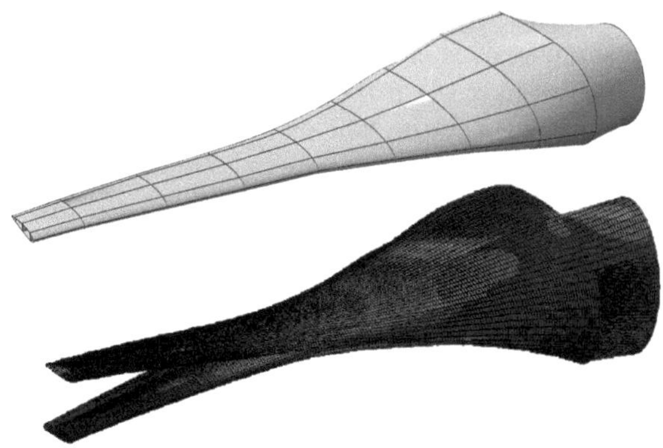

Fig. 7 Finite element model of a tidal blade (*top*) and flapwise deflection of the blade due to a tidal stream velocity

$$F_C = L \sin \theta + D \cos \theta, \tag{4}$$

where θ is the angle between the relative velocity and the plane of the rotor.

The input data for the HDM are water velocity, density, viscosity and initial velocity at the exit, number of blades, outer and inner diameter of the stream tube, RPM, C_L, and lift to drag ratio. Figure 6 shows an example of the radial distributions of F_A and F_C versus radius as predicted by the hydrodynamic model. Control systems have a significant effect on the forces and moments in a tidal turbine blade, by regulating the power absorbed by the turbine during high water velocity operation.

2.5 *Finite Element Analysis Model (Static)*

A finite element model (FEM) in conjunction with a fatigue life model (FLM) is used to characterise the strain distribution (i.e. maximum strain) in the turbine blade and estimate blade fatigue life. The example of the FEM modelling of tidal blade is given in Fig. 7.

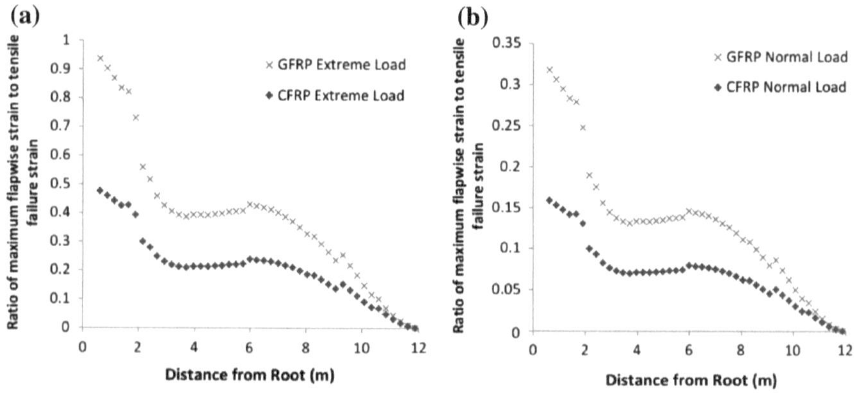

Fig. 8 Maximum flapwise strain in the blade's spar caps, using GFRP and CFRP, calculated using PreComp [34] for tidal stream velocities of **a** 4.1 m/s and **b** 2.4 m/s [9]

The maximum strains at each blade section are calculated [9] using the stiffnesses predicted from the software code PreComp [32] and the bending moment distributions predicted by the hydrodynamic model. The flapwise strains in the spar caps are significantly larger than the edgewise strains due to the large thrust forces on the blade. This static design approach can be used to compare the efficiency of using carbon fibre-reinforced polymers (CFRP) with the baseline GFRP. An example of predicted flapwise strain distribution along a 12 m length blade, for extreme and normal load cases, shows greater strain for the GFRP than for the CFRP spar caps (Fig. 8). The research shows that even when using thicker laminates (over 100 mm), the calculated strains are of the same magnitude as the failure strains of the material. Hence, the risk of failure over the lifetime of the turbine is significant [9]. The use of CFRP increases the static margin of safety in the blade, when compared to GFRP. This becomes important since high tidal velocities (up to 4 m/s) could occur and the effect of seawater degradation of the fibre matrix interface is likely to reduce the stiffness of the GFRP over the lifespan of such a device [8, 33]. Hence, the stiffness of the blade can be improved and the mass reduced by introducing a CFRP spar cap, for instance.

2.6 *Fatigue Loading of Blades*

The fatigue model focuses on the highest tensile strain in the blade, which is predominantly attributed to the flapwise bending-induced strain. Each cycle of the tide (2 per day) causes a slow increase and then a decrease in the maximum strain on the blade. The blade experiences a cyclic load for each revolution of the machine, caused by the blocking ('shadow') effect of the support tower (Fig. 9).

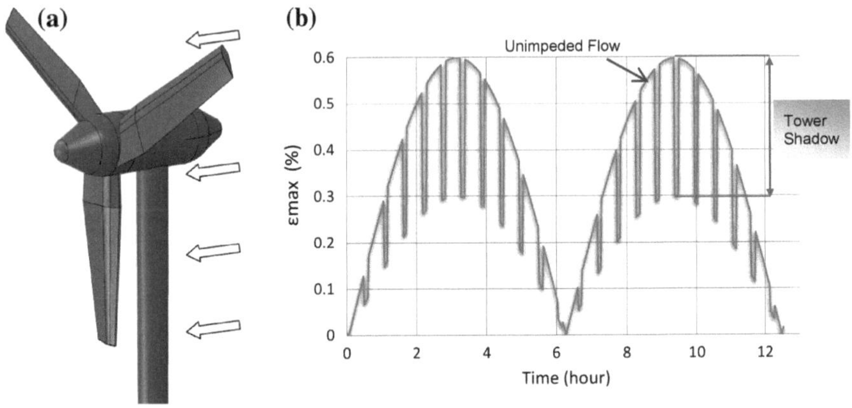

Fig. 9 **a** Tower shadow on downstream 3-bladed tidal turbine and **b** Schematic of tower shadow effect (50%) on maximum strain in the turbine blade

Experiments with scaled models of tidal turbines in a wave tank have found flapwise bending moment amplitudes of 50% [35]. This corresponds to a fatigue loading R-ratio ($\varepsilon_{Min}/\varepsilon_{Max}$) of 0.5 [24].

2.7 *Blade Fatigue Model*

The major strains in the blade are directly proportional to the flapwise bending moment, considering the material in the turbine blade to be essentially linearly elastic. Therefore, the maximum strain at any time *j* is

$$\varepsilon_{max,j} = \varepsilon_{max,sp}\left(\frac{M_f(v_j)}{M_{ref}}\right)^2, \tag{5}$$

where $M_f(v_j)$ represents the functional dependence of bending moment on tidal current velocity, M_{ref} the reference bending moment and ε_{ref} the maximum strain in the blade when the reference moment is applied corresponding to a specific reference velocity V_{ref}. The fatigue model has two components caused by two types of load (strain cycles). The first sums the damage due to the strain cycles caused by the low-frequency semi-diurnal tidal cycle, which implicitly includes the lower frequency 14-day tidal cycle effect, while the second is caused by the higher frequency cycles, due to the tower shadow effect.

3 Finite Element Analysis (Damage Model)

In the FE model of the blade, for a particular integration point, the damage fractions are summed to find the 7-day damage and hence the turbine life, using a Miner's rule approach [36]:

$$D_{7-day} = \sum\nolimits_{k,tide}^{N_{7-day}} \frac{1}{N_{f,k}} + \sum\nolimits_{j,rev}^{N_{7-day}} \frac{1}{N_{f,j}}, \quad (6)$$

where N_k is the number of tidal movements during the 7-day period, N_j is the number of turbine revolutions during the 7-day period, $N_{f,k}$ and $N_{f,j}$ are the numbers of cycles to failure for a given combination of mean and alternating strain during each tide and revolution, respectively, k is the increment of tidal cycles, and j is the increment of revolutions of the turbine. Hence, the blade life in years can be obtained as follows:

$$Blade\ life_{years} = \frac{7.38}{365.25 D_{7-day}}, \quad (7)$$

where 7.38 is the exact length of the 7-day period used (1/4 of a synodic month) and 365.25 the length of a year in days.

4 Manufacturing and Testing

The structural properties of GFRP materials depend on orientation of the fibres, polymer type and fibre/polymer volume fraction. One of the main advantages of fibre-reinforced materials is the ability to align the strong, stiff fibres with the main loads and thereby use the material to its maximum advantage. There are situations, however, particularly in relation to emerging technologies, e.g. marine renewable energy, where the loads are not very well understood and are complex and multidirectional in nature, for which quasi-isotropic (QI) laminates can be used.

The multidirectional nature of laminates complicates the fatigue damage mechanisms. Matrix cracking parallel to the fibres or inter-fibre failure (IFF) is first seen in the most off-axis plies under tensile loading. Ultimate IFF cracking takes place in the first 25% of fatigue life and the significant drop in stiffness is complete with only a minor reduction in stiffness after this point. However, fatigue strength reductions do not follow the changes in static strength since the damage mechanisms are different in fatigue [37].

4.1 Coupon Manufacture

The vacuum-assisted resin transfer moulding (VARTM) process was used to lay up 800 mm squares of $[(45°/135°/90°/0°)_2]_s$ laminates infused with epoxy or vinyl ester resin using (Fig. 10a).

The E-glass fabric used (Ahlstrom#42,007, Ahlstrom Corporation, Helsinki, Finland) is a biaxial non-crimp form with 2 plies (each 300 g/m^2) stitched together at 90°, while the 45°/135° plies were obtained by cutting an 800 mm square at a 45° angle biaxial fabric used for the 0°/90° plies. Eight layers of the fabric were stacked on the mould to create the 16 ply laminate. Resin feed tubes, vacuum tubes and spreader manifolds were placed on top of the stack of fabrics and a layer of breather fabric laid over the top of the entire surface to provide a path for the resin to spread across the laminate. A vacuum bag was laid over and sealed around the edges of the mould with a synthetic rubber sealant tape and the resin feed and vacuum tubing were brought out through the seal. The vinyl ester resin, Scott Bader VE676-03 (Scott Bader Company Ltd., Wellingborough, UK), was mixed with 2% Accelerator G, and then 2% Trignox 239 catalyst (Akzo Nobel N.V., Amsterdam, The Netherlands) was added and mixed thoroughly. The resulting mixture was allowed to stand for 5 min to degas before introduction into the fabric stack. It took approximately 25 min, at 18 °C room temperature, to infuse into the 0.8 m 0.8 m × 3.75 mm laminate. The process was repeated to manufacture the

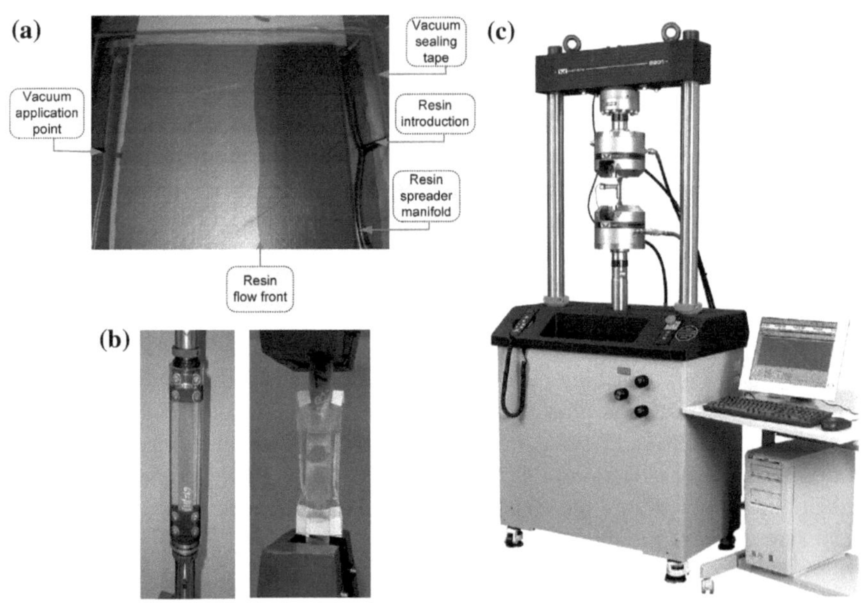

Fig. 10 Coupon manufacturing and testing. **a** Vacuum-assisted resin transfer moulding process; **b** Saturated specimens immersed in tap water (using acrylic tube with O-ring seal and polythene pouch with waterproof tape) during testing; and **c** Instron 8801 fatigue test machine

epoxy/E-glass laminates (the hand lamination formula of Ampreg 22 with slow hardener—Gurit Holding AG, Wattwil, Switzerland). Since this resin is more viscous than is optimum for resin infusion, the viscosity was lowered by increasing the temperature to 27 °C and it took an hour to infuse into the laminate.

The vacuum continued to be applied for a number of hours after infusion, until the resin had set and after a room temperature cure for 48 h, the laminates were oven cured at 80 °C for approximately 4 h. Fibre volume fractions of 50% were achieved at an average thickness of 3.75 mm. The laminates were cut into 25 × 250 mm coupons (Fig. 10b) using a water cooled, diamond-tipped, rotary saw under controlled feed rate.

4.2 Fatigue Testing

An Instron 8001 test machine with 8800 controller was used for fatigue testing (Fig. 10b, c), applying a sinusoidal force–time history in tension–tension mode ($R = 0.1$), at frequencies of between 3 and 6 Hz. All static tests and a small number of fatigue tests were carried out on an Instron 8500 test machine with an 8800 controller. The gripping pressure on both machines was adjusted to 50 MPa and 100 MPa for fatigue and static tests, respectively. The fatigue test frequency was decreased to maintain a constant rate of strain application as the strain level increases ensuring that the coupon does not overheat. A fan was also used to cool the coupon during testing. A radiation thermometer measured less than 10 °C temperature rise in surface temperature during the highest stress cycling test. The temperature in the test room with the 8001 machine was maintained at 21 °C and with the 8500 machine varied between 15 and 19 °C. Five coupons were tested at each of four separate stress levels (Fig. 10c). Approximately, twenty coupons of each material were tested over a range of maximum stress levels that caused the coupons to fail between 100 and 10 million cycles. The number of cycles to failure and the maximum fatigue strain applied to each coupon was recorded, producing a strain–life (ε-N) curve. A power law relationship between maximum initial strain ($\varepsilon_{max,i}$ and fatigue life is

$$\varepsilon_{max,i} = AN_f^{-B}, \tag{8}$$

where N_f is the number of cycles to failure, and A and B are constants for power law fit to fatigue data.

Vinyl ester/E-glass has a number of advantages over the epoxy/E-glass material used, e.g. lower cost and increased resistance to water penetration. However, the epoxy resin shows superior fatigue performance over the vinyl ester resin. The strain–life fatigue test results for the epoxy/E-glass and vinyl ester/E-glass materials along with the corresponding power law curve fits, Eq. (8), are presented in Fig. 11. During the testing, epoxy proved to be more tolerant to the cracks and minor delaminations (appear at the edges of the coupon during testing). Furthermore,

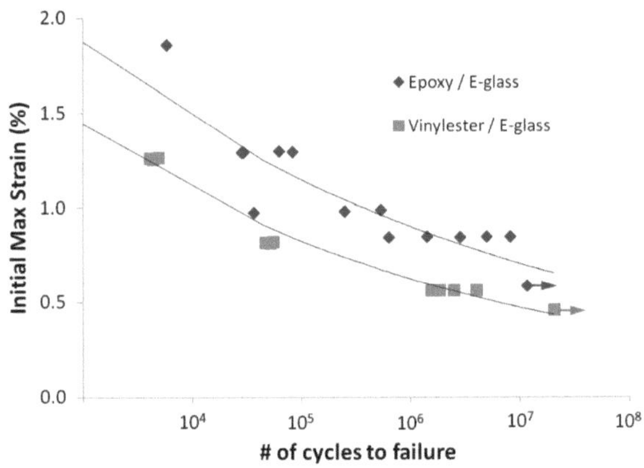

Fig. 11 Test results for Quasi-Isotropic (QI) laminates (R = 0.1)

Table 1 The list of the materials tested and ageing procedures used

Material	V_f	Thickness (mm)	Immersion ageing (mths)	Stress aged (mths)
Vinyl ester/E-glass	50% ± 2%	3.75	40 °C/30	21
Vinyl ester/E-CR glass	60% ± 2%	4.22	40 °C/20	
Epoxy/E-glass	50% ± 2%	3.75	30 °C/29	

applying experimental fatigue data for both of the materials in the fatigue model and integrating with respect to time history of tidal velocity, it is predicted that the maximum strain in a vinyl ester blade must be 29% less than in an epoxy blade, to achieve the same fatigue life.

5 Effects of Water Saturation

GFRP tends to absorb a small amount of water, typically less than 5% [12]. The water diffuses into the polymer matrix, changing its mechanical properties. The water fills any small voids and tends to dilute any unreacted constituents of the polymer, while the resulting osmotic pressure can cause damage to the laminate, i.e. stress corrosion cracking (SCC) of the fibres may occur.

The materials and processes have been chosen to create state-of-the-art GFRP laminates in order to perform accelerated ageing on these laminates at a low temperature, and testing in order to reliably mimic the water saturation level found in an operating tidal turbine blade as described in Sect. 4.1.

5.1 Accelerated Ageing Procedure

Accelerated ageing was used to simulate the effect of approximately 20 years in cold seawater by immersing the QI specimens, with either epoxy or vinyl ester

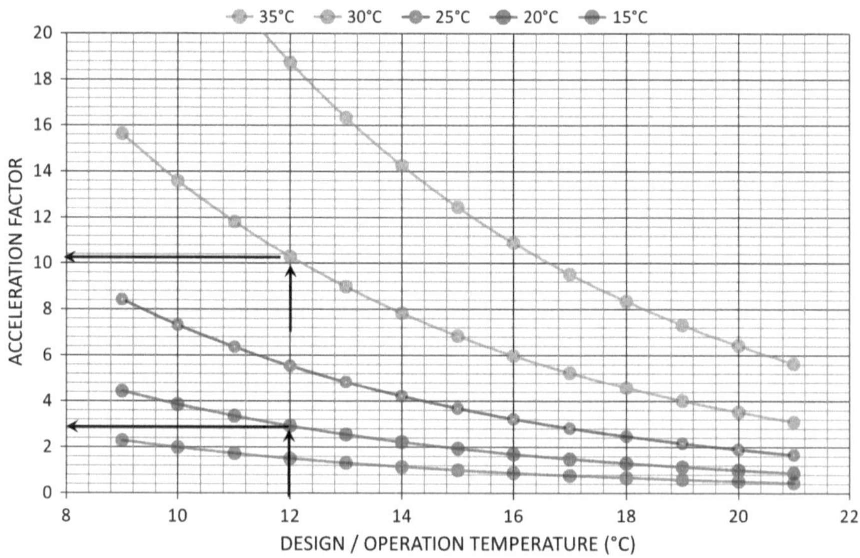

Fig. 12 Predicted acceleration factor for epoxy/E-glass composite due to ageing water temperature

matrix reinforced with E-glass or E-CR glass (Table 1), in warm (30–40 °C) water for up to 2.5 years [13]. Movement of moisture by diffusion into the polymer resin and damage processes within the composite was accelerated by increasing the temperature of the material.

Figure 12 shows a plot of the acceleration factors for epoxy with the diffusion constants that result from combinations of operating (reference) temperatures and accelerated ageing temperatures [38]. 12 °C has been chosen as the design operating temperature for a tidal turbine operating off the coast of Ireland. The graph shows that the 20 and 30 °C ageing temperatures will cause diffusion-controlled processes in epoxy/E-glass composites to proceed approximately 3 and 10 times faster, respectively, than at the 12 °C operating environment. The water temperature for vinyl ester ageing was increased to 40 °C, which results in an acceleration factor of 10 [13, 38]. Hence, immersing vinyl ester glass composite in a water bath at 40 °C for one year will cause damage equivalent to 10 years in 12 °C seawater.

Two tap water baths with heater stirrers (heated water is distributed throughout the entire tank) were used during the specimen age-immersion process. One tank contained the epoxy/E-glass coupons and was maintained at 30 °C. The other heated tank contained vinyl ester/E-glass coupons and is maintained at 40 °C. A third unheated tank is used to age both the epoxy and vinyl ester composites at room temperature. The maximum water circulation around the coupons was allowed while they were supported evenly, so that there were no unwanted stresses generated in specimens during the saturation. Each of the coupons was weighed on a mass balance and the weight recorded before it was placed in the tank. The average weights for the Epoxy/E-glass and Vinyl ester/E-glass coupons were 43.872 g (4.1% variation) and 43.596 g (1.4% variation), respectively. The coupons were reweighed before testing (after the immersion ageing period) to establish the amount of water that had been absorbed during ageing.

5.2 *Stressed Immersed Testing*

Figure 13a shows a vinyl ester/E-glass coupon installed in a stressed test configuration. The main component of this rig is a heavy duty die spring (Berger R50-203, Berger Tools Ltd., Kemsing, Kent, UK), with a spring rate of 117 N/mm. The end cap of the stressing mechanism transfers the load from the die spring end into stainless steel pin. The end cap has a slot machined in it to allow a 25 × 4 mm coupon to pass through. The coupon was modified slightly to work with the mechanism by having an 8–mm-hole drilled on each end. During assembly of the mechanism the spring was compressed by 34.2 mm and trapped between the end caps and pins inserted in the holes in the coupon, thereby applying a 4 kN load to the composite coupon. This resulted in a stress of 42.7 MPa and a strain of 0.21% in the 25 × 3.75 mm QI vinyl ester/E-glass coupons. Before each of the ten coupons is assembled into the stressing rig, they were fatigued in air for 10,000 cycles between 0.8 kN and 8 kN. This load level was chosen to cause cracking in the 90° plies of the material in order to mimic the damage expected to occur in service. Five of the stressed coupons were immersed in 40 °C water for 20 months, while the other five were stored in insulated conditions at the same temperature as the water bath. The fatigue testing setup is shown in Fig. 13b while the testing procedure is described in Sect. 4.2.

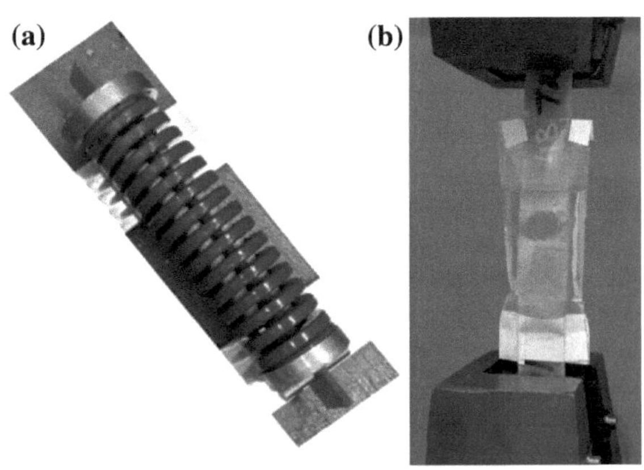

Fig. 13 **a** Stressing mechanism for stress ageing of GFRP coupon and **b** Immersed fatigue test setup, polyethylene pouch with waterproof tape

5.3 Water Uptake During Immersion Ageing

The rate of water absorption at 30 °C and 20 °C (room temperature) by the epoxy/E-glass coupons is shown in Fig. 14a. The results show that the coupons immersed in the 30 °C ageing tank absorbed water more quickly than those at room temperature, becoming saturated at 0.6% moisture. This is expected as the higher temperature in the ageing tank triggers the higher diffusion of the water molecules. On the contrary, the coupons at room temperature had not become saturated by the end of the immersion period. The results of the vinyl ester/E-glass coupons immersion show the same trend (Fig. 14b). The vinyl ester/E-glass coupons immersed in 40 °C water for 21 months on average absorbed 0.16% of their weight. As previously observed, the lower immersion temperatures resulted in slower saturation of the specimens.

The amounts of water absorption by the epoxy and vinyl ester composite coupons are significantly lower than previously reported in the literature [39, 40]. There are two possible reasons why this phenomenon occurred: (a) the particular polymers chosen for this work are for use in immersed applications, thus they have been developed to minimise water absorption, and (b) the laminates from which the coupons were taken had a low void content.

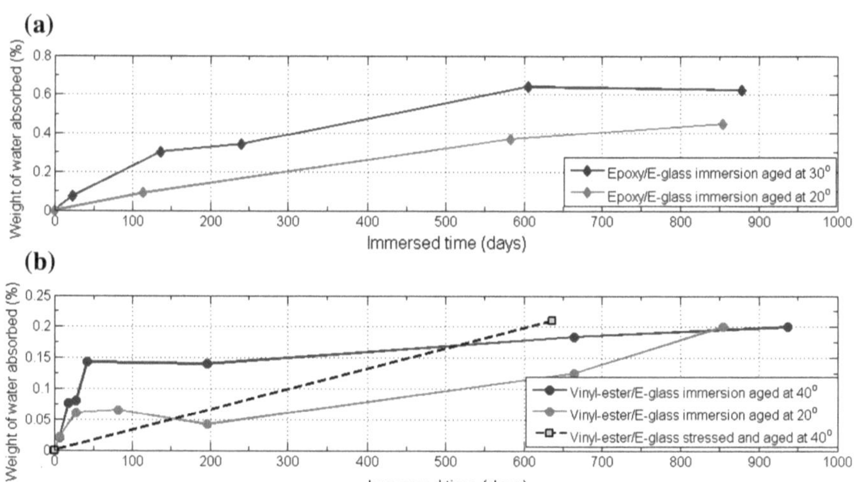

Fig. 14 Weight of water absorbed by **a** Epoxy/E-glass and **b** Vinyl ester/E-glass composite during water immersion ageing

5.4 *Effects of Water Saturation on Fatigue Life of E-glass/Epoxy*

Constant amplitude fatigue testing of both dry and wet coupons established the stress–life curves for the materials and thereby quantified the degradation in the fatigue strength due to water saturation of the materials. The results of fatigue testing epoxy/E-glass QI coupons in tension–tension mode ($R = 0.1$) for the wet and dry coupons are shown in Fig. 15. The dry coupons were stored at room temperature for 20 months and tested in air at room temperature, while the wet coupons were immersed for 20 months in 30 °C water and tested while immersed, in the water pouch (Fig. 13b), at room temperature. The results show a drop in fatigue strength for the water aged and immersed coupons compared to the coupons which were aged and tested in normal room temperature air. At high stresses the wet aged fatigue strength decreases by 20–25% and at high cycles the wet aged strength is 8% below the fatigue strength of the dry material. The actual change in fatigue life of a structure will depend on the spectrum of fatigue cycles loads it experiences while it is in service. It is observed that during fatigue testing all the immersed coupons broke in the zone that was actually immersed in water within the pouch which indicates that it is important to keep the coupons wet during testing for best representation of the immersed failure condition, as any drying that takes place during the testing tends to be accompanied by a recovery in strength.

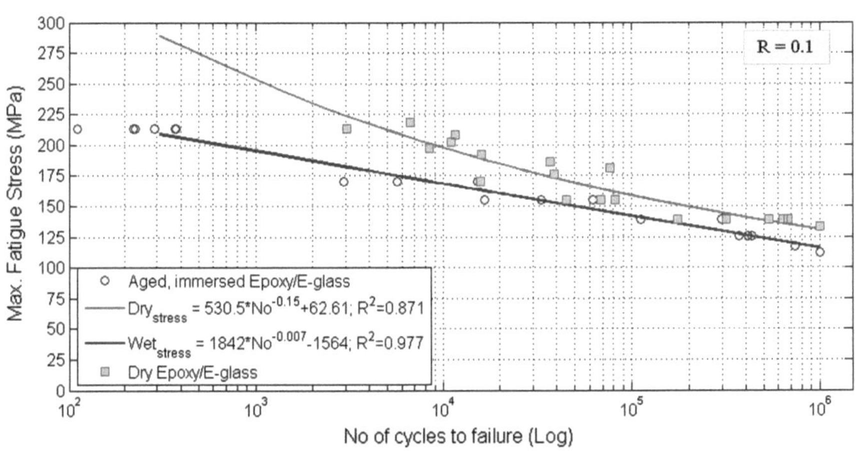

Fig. 15 Stress–life curves for wet and dry epoxy/E-glass in R = 0.1 fatigue tests

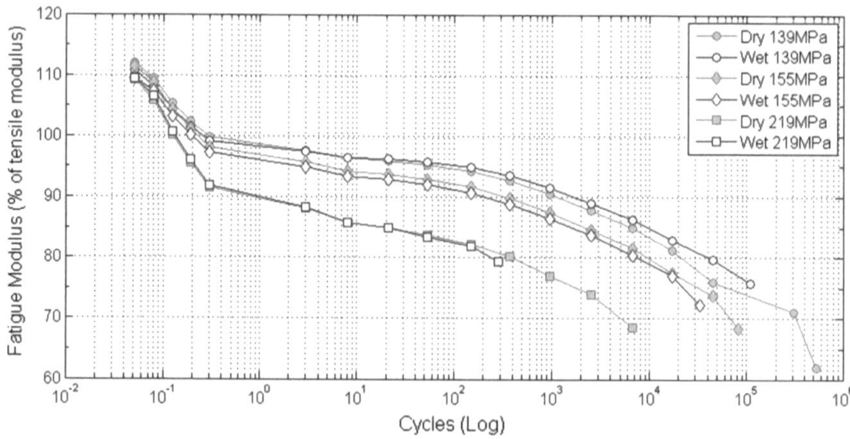

Fig. 16 QI epoxy/E-glass Modulus degradation during fatigue cycling at R = 0.1

5.5 *Fatigue Modulus Reduction in Wet and Dry GF/Epoxy*

The fatigue modulus is monitored during the fatigue tests on the dry and wet coupons. The evolution in fatigue modulus over the life of wet and dry coupons at three different load levels shows that wet coupon stiffness decays in the same way as the dry coupons during fatigue cycling (Fig. 16). The only significant difference between the behaviour of the wet and dry coupons during fatigue cycling is that the wet coupons consistently fail at a lower number of cycles.

6 Tidal Turbine Blade Design Example

6.1 *Tidal Blade Model*

The following design example is based on 3-bladed downstream, free-yaw tidal device with a 5.0 m rotor radius, operating in a 2.6 m/s tidal current velocity, and with approximately 330 kW production capacity. The input parameters for the HDM (generated by tidal model) are shown in Table 2.

6.2 *Hydrodynamic Model*

The algorithm of the HDM, as described in Sect. 2.4, is shown in Fig. 17. The design of the aerofoil (chord and twist distribution along the blade) is found so that the turbine has the optimum performance over its lifetime. The blade is divided into

Table 2 Parameters for base case hydrodynamic model

Parameter	Value
Water velocity	2.6 m/s
Number of blades	3
R_{outer}	5.0 m
RPM	16
C_L	1.0
L/D	70
Angle of attack	7°
ζ	0.333
Water density	1025 kg/m^3
Water viscosity	0.0,013,155 Pa/s
R_{inner}	1.5 m

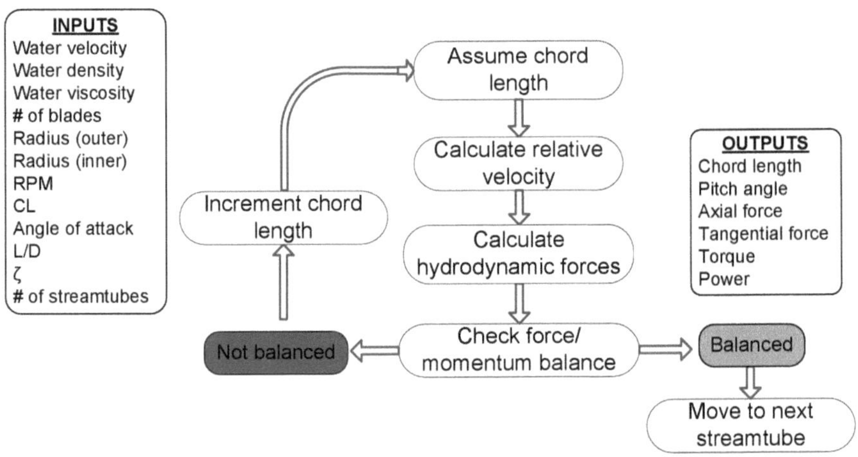

Fig. 17 Summary of hydrodynamic model method

number of sections (stream tubes). The HDM performs adjustments of the chord length until the moment balance is achieved, after which the process is repeated for all remaining stream tubes. For the set of input parameters (Table 2) model generates the radial distribution of blade chord length, pitch angle, tangential and axial forces, and power. In this example, the chord length at the blade root is 1.25 m with the pitch angle 20°, while the chord length at the tip is 0.75 m with the pitch angle 4°. F_A increases linearly with increasing radius, from 0.7 to 2.3 kN at the extreme blade radius, and is significantly larger than the (power-generating) F_C, which increases with the radius from 320 N to 410 N at the extreme blade radius (Fig. 6). These forces cause a flapwise bending moment of 150 kNm and edgewise bending moment of 35.7 kNm, respectively, at a 1.5 m radius from the rotor centre.

6.2.1 Pitch-Regulated Tidal Turbines

In order to regulate the turbine power during high water velocity, control systems are used to manage forces and moments on the tidal blade. The HDM is used to simulate the two options for controlling power in tidal turbines, pitch-regulated (PR) and stall-regulated (SR). PR is a system which modifies the lift coefficient (C_L), i.e. the forces on the blade, by rotating the entire blade about its axis. SR blades are designed with a radial variation of pitch angle so that the angle of attack over a section of the blade exceeds the stall angle and lift drops off, reducing the forces on the blade, with tidal velocity increase [41]. An operational range of velocity for tidal currents is typically 0.75 to 3 m/s (exceptionally 4 m/s). For this example the operational environment is such that both control systems produce the same flapwise moment [42].

The maximum theoretical power levels for SR and PR turbines and the energy produced by each turbine per year are calculated using the HDM [24]. The energy output of the PR turbine is matched to that of the SR turbine to enable an objective comparison with the respect to damage accumulation. This is achieved by identifying the threshold value of water velocity for pitch control, above which blade pitch is controlled to give constant power. Figure 18a, b shows the power and thrust moment curve for a PR and SR tidal turbine, respectively. These turbines would have similar energy output in a year at 3.05 m/s current speed.

6.2.2 Effect of Control Strategy on Bending Moments

The maximum fibre strain at any water velocity, given the flapwise bending moment at that water velocity and the assumption of linear elasticity in the blade, can be computed using Eq. (5) given in Sect. 2.7. The predicted dependence of bending moment near the root of the blade on water velocity, due to thrust forces calculated by the HDM, for a PR, an SR and an unregulated tidal turbine is shown

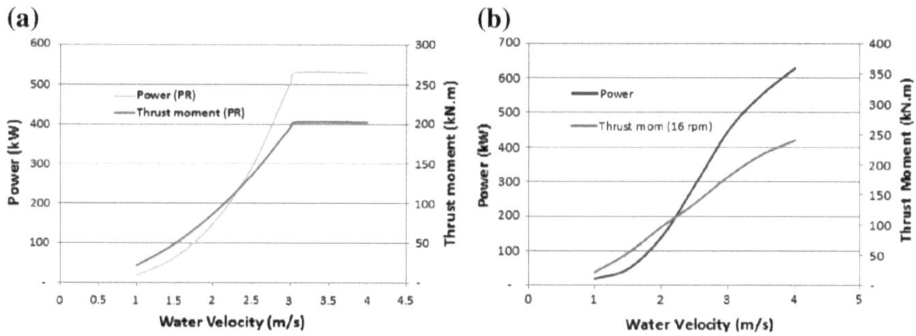

Fig. 18 Power and thrust moment of **a** pitch- and **b** stall-regulated tidal turbine operating in different flow regimes

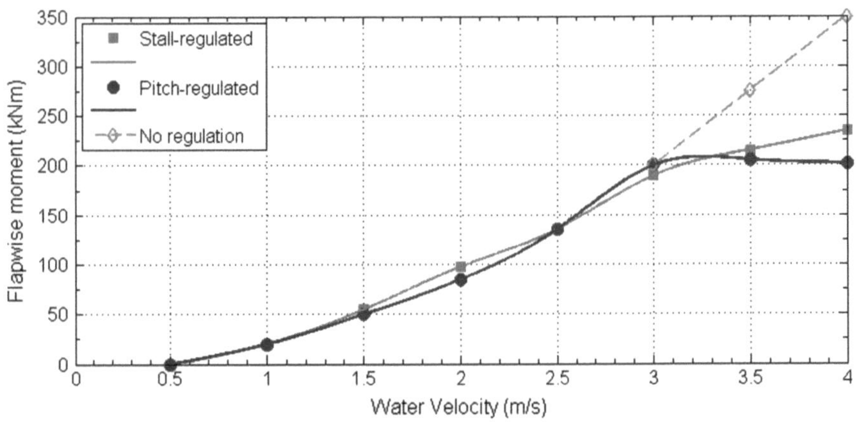

Fig. 19 Predicted effect of water velocity on blade bending moment (at R = 1.5 m) for various control strategies, as calculated using hydrodynamic model

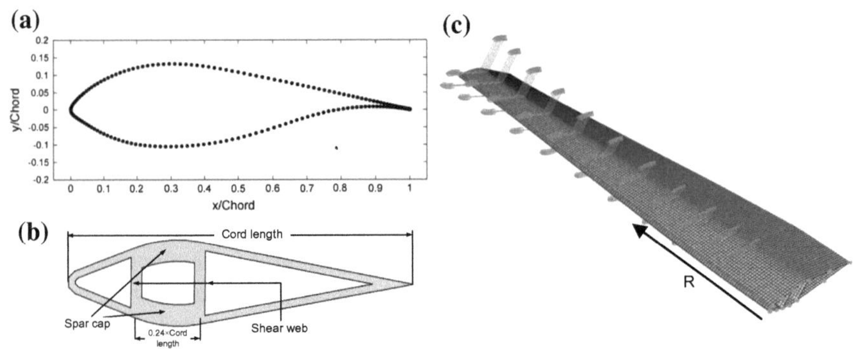

Fig. 20 **a** RIS–A1-24 Airfoil [45]; **b** Simplified aerofoil shape used for finite element analysis of a tidal turbine blade and **c** Final element (FE) model of a 5 m blade for a 3-bladed, 330 kW turbine

in Fig. 19. The turbines operate in the water velocity regime where all three control systems produce the same flapwise moment (Sect. 6.3).

6.3 *Structural /Finite Element Model*

The aerofoil shape has been simplified (Fig. 20a) for structural modelling, by a piecewise linear equivalent keeping the same section thickness, 24% of chord length (Fig. 20b). The two shear webs have been inserted into the aerofoil shape creating a box section which becomes the structural core of the blade. The top caps and shear webs of the box are 35 and 12 mm thick near the root (at 1.5 m radius)

Table 3 Epoxy/E-glass material properties

Inputs		Single unidirectional ply properties		Quasi-isotropic laminate [(45\135\90\0)$_2$]$_s$	
V_f	50%	$\mathbf{E_1}$	38 GPa	$\mathbf{E_x, E_y}$	19.3 GPa
E_f	72.4 GPa	$\mathbf{E_2}$	11.6 GPa	$\mathbf{G_{xy}}$	7.2 GPa
ν_f	0.22	$\mathbf{G_{12}}$	3.5 GPa	$\boldsymbol{\nu_{xy}}$	0.330
E_m	3.5 GPa	$\boldsymbol{\nu_{12}}$	0.285		
ν_m	0.35				

Fig. 21 Undeformed and deformed shape of blade at 2.6 m/s water velocity, with a factor of safety of 2.0 applied

Table 4 Parameters for fatigue reference case

Parameter	Value
Maximum water velocity	4.0 m/s
Neap max. velocity/Spring max. velocity	60%
Factor of safety applied to loads	2
Tower shadow	50%
Control system	PR
Material	Epoxy/E-glass
Spar top cap thickness range	35–6 mm
Shear web and fairings thickness range	12–4 mm

and taper to 6 and 4 mm at the tip, respectively. A finite element (FE) model of a tidal turbine blade has been created using Abaqus FEA software [43] (Fig. 20c) with the panels of the blade modelled as shell elements with QI material properties from standard laminate analysis [44] (Table 3).

The deflected shape of the blade for the extreme loading, taking into account a safety factor of 2, applied to the FE model is shown in Fig. 21. Under the factored load the blade tip deflects 334 mm axially, which is approximately twice the blade aerofoil thickness at the tip. This amount of deflection will not introduce any significant error into either the structural or hydrodynamic calculations [24].

In order to predict fatigue damage accumulation, using equations given in Sects. 2–4, the moment–velocity relationships (Fig. 19) are fitted with polynomial expressions, to allow interpolation with respect to water velocity. The fatigue reference case (Table 4) is analysed by the fatigue model which predicted a fatigue life of 11.6 years for the observed case scenario. The fatigue model (Sect. 2.7) can give insight into the effects of each of the parameters shown in Table 4, on the fatigue life.

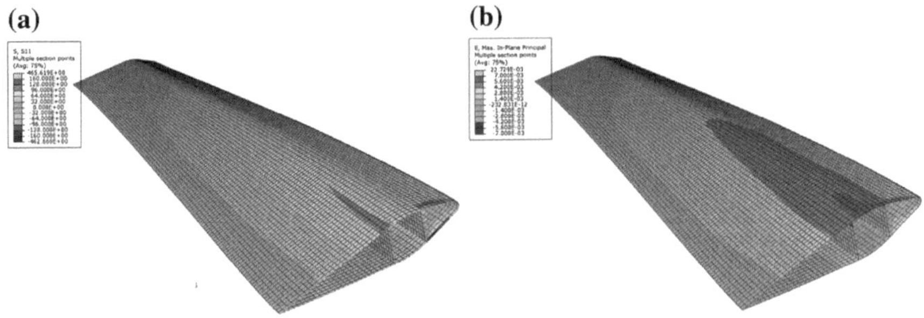

Fig. 22 Final element model results; **a** First principal stress distribution in 45° surface ply under design load and **b** Maximum principal strains in the outer surface along the blade under design load

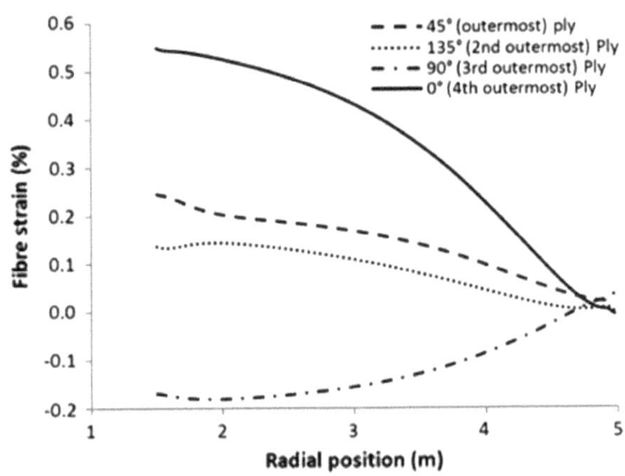

Fig. 23 Fibre strains in the four outermost plies at maximum strain locations along blade operating in 2.6 m/s water velocity

The results of the FE analysis show that the highest stresses and strains occur near the blade root, in the spar caps of the structural beam box section (Fig. 22). These are primarily caused by the large (flapwise) bending moments due to the thrust force, F_A (Fig. 6).

The maximum bending strains occur on the outer surface, while the maximum principal strains are in the middle of the spar cap. In order to make an appropriate comparison to the experimental fatigue life data, it is necessary to find the maximum fibre strain in a ply in which the fibres are closely aligned to the load direction (note that surface ply is 45° to the blade axis). Fibre strains along a section through the midline of the pressure side spar cap extending from near the hub to the tip of the blade are shown in Fig. 23. The fibres in the three outermost plies are at 45°, 135° and 90° to the blade axis (i.e. to the maximum principal strain) which reduces their fibre strains. The 4th ply from the surface has its fibres at 0° to the blade axis which means that is aligned with the direction of the principal bending strains. The maximum fibre strains are predicted in this ply (Fig. 23).

The base case results have given the maximum fibre strain in the turbine blade at 2.6 m/s water velocity. To compute maximum fibre strain at any water velocity, given the flapwise bending moment at that water velocity with the assumption of linear elasticity in the blade, Eq. (5) can be employed.

6.4 Composite Damage (Pitch vs Stall Regulation)

The blade life of SR and PR devices is determined using the fatigue model. Based on the water velocity increments of 0.1 m/s the fatigue damage fraction for each rotation of the turbine is calculated. The results of cumulative damage for the PR and SR blades for different velocities are shown in Fig. 24. The area under each individual curve represents the normalised total accumulated damage over the blade's lifetime.

The bulk of the damage that occurs in the SR blade is at high water velocities (greater than 3.5 m/s). On the contrary, the PR blade experiences more damage at low water velocities. For example, the damage fractions of SR and PR turbine blades at 3.9 m/s water velocity are 0.149 and 0.066, respectively. The reason for the SR turbine blade's behaviour at such high velocities is the relatively higher bending strain in comparison with the PR blade. However, the PR blade experiences the most of the damage at medium water velocities (2.5–3.5 m/s). This is consistent with the fact that PR blades predominately operate in this water regime. Hence, when comparing the blades with the same structural design, employed on two differently controlled tidal turbine devices it was found that the SR blade has a shorter life span (12.8 years) than PR blade (20 years).

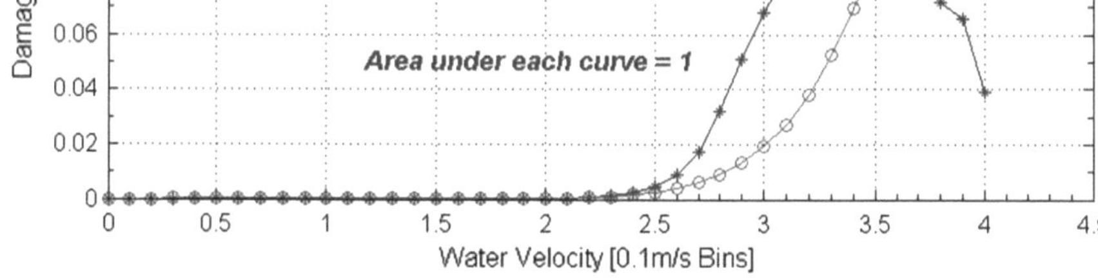

Fig. 24 Damage accumulation per 0.1 m/s bin for PR and SR epoxy/E-glass turbine blades operating in a tidal flow with 4.0 m/s maximum velocity and 60% neap/spring variation

7 Effect of Water Saturation on Predicted Blade Fatigue Life

The predicted fatigue life for a blade on an SR and PR tidal turbine is plotted against maximum stress experienced by the blade in a 2.5 m/s tidal flow (Fig. 25). The predicted 20-year "dry" fatigue life of a stall-regulated tidal turbine blade is reduced to 17 years if the laminate is water saturated. For a 66 MPa maximum stress in the blade, the predicted "dry" life is 5 years and the predicted "wet" life is 2.9 years. Kennedy et al. have shown that a stall-regulated blade with "dry" laminates needs only 10% thicker laminates in order to have equal life with a pitch-regulated blade [24]. However, the analysis shows that the laminates would need to be an additional 1 to 4% thicker to counteract the effects of water saturation on the blade. A blade of PR turbine with a maximum stress of 72 MPa, which is predicted to have a 5 year "dry" life, will fail, on average, 1.7 years earlier if the blade laminates are water saturated. However, the "dry" life of 20 years and the "wet" life of 19.1 years when water saturated can be achieved by increasing the thickness of the laminates, and thus reducing the stress in the blade to 63.5 MPa. This convergence of the wet and dry laminate lives occurs because pitch regulation limits the maximum stress in the blades. In achieving the longer blade fatigue life (> 15 years), stresses are limited to levels where the difference in life between wet and dry laminates is very small. According to the results shown in Fig. 25, the life of water saturated blades is reduced by about 3 years for SR and about 1–2 years for PR turbines.

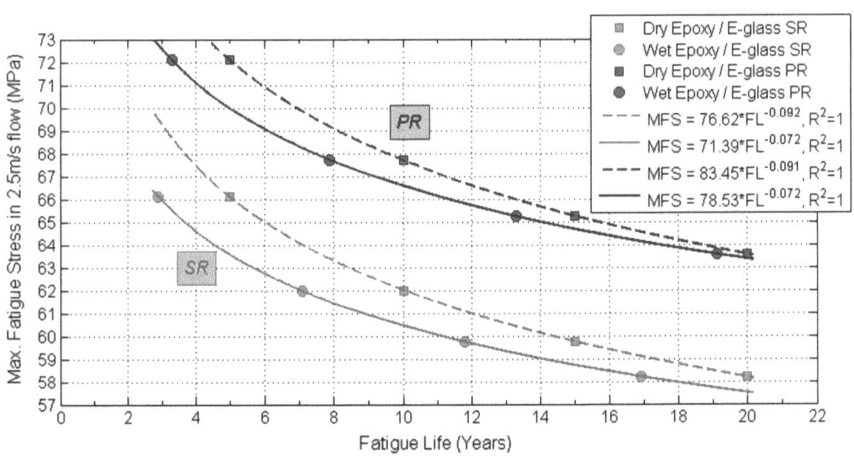

Fig. 25 Effect of seawater saturation on predicted fatigue life of pitch-regulated (PR) and stall-regulated (SR) tidal turbine blades

8 Conclusions

An approach for the fundamental fatigue design of glass fibre-reinforced polymer (GFRP) tidal turbine blades is described in this study. The model is implemented with the aim to estimate and compare the immersed life of stall-regulated (SR) and pitch-regulated (PR) tidal turbine blades. In that regard, the uniaxial fatigue testing results of water-immersed Quasi-Isotropic (QI) GFRP, i.e. epoxy/E-glass and vinyl ester/E-glass materials, are incorporated in the three-bladed tidal turbine model for a range of tidal velocities.

The mechanical properties of the material used in the blades are evaluated. There was only 1% drop in QI epoxy/E-glass coupons modulus when immersed in 30 °C for 20 months. However, there was no difference between the fatigue modulus for the wet and dry specimens. For the increased water temperature the water absorption was relatively low, i.e. for epoxy/E-glass (0.6% for 20 months at 30 °C) and vinyl ester/E-glass (0.2% for 22 months at 40 °C). Regardless of the low level of water absorption, the fatigue strength of the material was significantly degraded by water immersion ageing. It was found that the wet fatigue strength is 20% and 8% lower than the dry fatigue strength at 1000 cycles and at 1 million cycles, respectively. Hence, the effect of water saturation is also stress level dependent. The stressing of the coupons during immersion did not result in additional water absorption. However, these specimens experienced 10% loss in fatigue strength. Overall, it is found that the epoxy/E-glass material is significantly more fatigue resistant than the vinyl ester/E-glass material and thus more desirable in tidal blade design.

Nonetheless, the experiments showed that the fatigue life of tidal blades is extensively dependant on strain–stress level experienced by the blade, regardless of the material. For instance, an increase in a material stress of only 15% could decrease blade life by up to 75%. Likewise, the tower shadow effect influences the fatigue life of blades, so that increasing the shadow effect from 50% to 90% will lead to 15% reduction in the blade fatigue strength or maximum allowable strain. Regarding tidal turbine blade life span, the study found that an SR blade will have a shorter life than an equivalent PR turbine blade. In order to increase the fatigue blade life of the SR blade to that of PR turbine blade, the thickness of the laminates in the spar cap needs to be increased by approximately 10%, i.e. decreasing the maximum strain by about 5%. However, in the case of both devices, the laminates in the spar need to be between 1% and 4% thicker in order to prevent the negative effect of water saturation on the blade.

Acknowledgements The authors would like to acknowledge funding from Science Foundation Ireland (SFI) through the Advance Award (14/ADV/RC3022) and the Marine and Renewable Energy Ireland (MaREI) Centre, Grant No. 12/RC/2305; and from the European Union under Framework 7, through the MARINCOMP Project, Grant agreement no.: FP7-612,531.

References

1. NASA (2016) TOPEX/Poseidon: revealing hidden tidal energy. Retrieved 14.10.2016, 2016, from http://svs.gsfc.nasa.gov/stories/topex/tides.html
2. UKCE (2012) UK wave and tidal key resource areas project summary report the crown estate. Retrieved from https://www.thecrownestate.co.uk/media/5476/uk-wave-and-tidal-key-resource-areas-project.pdf
3. SEI (2006) Tidal and current energy resources in ireland report. Dublin, Ireland
4. Magagna D, Uihlein A (2015) Ocean energy development in Europe: Current status and future perspectives. Int J Mar Energ 11:84–104. doi:10.1016/j.ijome.2015.05.001
5. O'Rourke F, Boyle F, Reynolds A (2010) Tidal current energy resource assessment in Ireland: Current status and future update. Renew Sustain Energy Rev 14(9):3206–3212
6. Little C, Grimwade J, Court RS (2012) Recommendations on the structural testing requirements for tidal stream blades paper presented at the ICOE2012, Dublin, Ireland. https://www.icoe-conference.com/publication/recommendations_on_the_structural_testing_requirements_for_tidal_stream_blades/
7. Jaksic V, Kennedy CR, Leen SB, Brádaigh CMÓ (2016, 21–22 March) Tidal turbine blade design from a fatigue point of view. In: paper presented at the oxford tidal energy workshop, OTE2016, department of engineering science, Oxford, UK
8. Davies P, Rajapakse YDS (2014) Durability of composite in marine environment (vol. 208): Springer
9. Grogan D, Leen S, Kennedy C, Brádaigh CO (2013) Design of composite tidal turbine blades. Renew Energ 57:151–162
10. Fagan EM, Kennedy CR, Leen SB, Goggins J (2016) Damage mechanics based design methodology for tidal current turbine composite blades. Renew Energ 97:358–372. doi:10.1016/j.renene.2016.05.093
11. MARINCOMP: EU FP 7 Marie Curie IAPP Project MARINCOMP—"Novel Composite Materials and Processes for Marine Renewable Energy", Grant agreement number: FP7-612531, Funded under: FP7-People. www.marincomp.eu
12. Baba NB, Suhaimi AS, Mohd Amin MA, Mohd A (2015) Study on mechanical and physical behaviour of hybrid GFRP. Adv Mater Sci Eng 2015:7. doi:10.1155/2015/138965
13. Kennedy CR, Leen SB, Ó Brádaigh CM (2016) Immersed fatigue performance of glass fibre-reinforced composites for tidal turbine blade applications. J Bio-and Tribo-Corrosion 2 (2):1–10. doi:10.1007/s40735-016-0038-z
14. Degrieck J, Van Paepegem W (2001) Fatigue damage modeling of fibre-reinforced composite materials: review. Appl Mech Rev 54(4):279–300
15. DNV GL—Energy (2014) Horizontal axis tidal turbines Renewables Certification—Wave and Tidal (pp. 284)
16. DNVGL (2015) Tidal turbines—Rules and standards
17. Vassilopoulos AP, Manshadi BD, Keller T (2010) Influence of the constant life diagram formulation on the fatigue life prediction of composite materials. Int J Fatigue 32(4):659–669. doi:10.1016/j.ijfatigue.2009.09.008
18. Duan X, Yao WX (2002) Multi-directional stiffness degradation induced by matrix cracking in composite laminates. Int J Fatigue 24(2–4):119–125. doi:10.1016/S0142-1123(01)00066-4
19. Talreja R (2013) Continuum modeling of the development of intralaminar cracking in composite laminates. In: Paper presented at the ICF7, Houston (USA) 1989
20. McCann GN, Rawlinson-Smith RI, Argyriadis K (2006) Load simulation for tidal turbines using wind turbine experience. In: Paper presented at the ICOE'06, Bremerhaven, Germany
21. Bahaj AS, Batten WMJ, McCann G (2007) Experimental verifications of numerical predictions for the hydrodynamic performance of horizontal axis marine current turbines. Renew Energ 32(15):2479–2490. doi:10.1016/j.renene.2007.10.001

22. Batten W, Bahaj A, Molland A, Chaplin J, Group SER (2007) Experimentally validated numerical method for the hydrodynamic design of horizontal axis tidal turbines. Ocean Eng 34(7):1013–1020
23. Bahaj A, Molland A, Chaplin J, Batten W (2007) Power and thrust measurements of marine current turbines under various hydrodynamic flow conditions in a cavitation tunnel and a towing tank. Renew Energ 32(3):407–426
24. Kennedy CR, Leen SB, Brádaigh CMÓ (2012) A preliminary design methodology for fatigue life prediction of polymer composites for tidal turbine blades. Proc Instit Mech Eng, Part L: J Mater Design Appl, 1464420712443330
25. Jaksic V, Kennedy CR, Leen SB, Brádaigh CMÓ (2016, 21–22 March) Tidal turbine blade design from a composite fatigue point of view. In: Paper presented at the 5th oxford tidal energy workshop, Oxford, UK
26. Bryans AG, Fox B, Crossley PA, O'Malley M (2005) Impact of tidal generation on power system operation in Ireland. IEEE Trans Power Syst 20(4):2034–2040
27. Pugh DT (1988) Tides, Surges, and Mean Sea-Level/a Handbook for Engineers and Scientists. Wiley
28. Tidal Observations—Irish National Tide Gauge Network (2016) from Marine Institute http://www.marine.ie/Home/site-area/data-services/real-time-observations/tidal-observations
29. Batten WMJ, Bahaj AS, Molland AF, Chaplin JR (2008) The prediction of the hydrodynamic performance of marine current turbines. Renew Energy 33(5):1085–1096. doi:10.1016/j.renene.2007.05.043
30. Batten WMJ, Bahaj AS, Molland AF, Chaplin JR (2006) Hydrodynamics of marine current turbines. Renew Energy 31(2):249–256. doi:10.1016/j.renene.2005.08.020
31. UIUC (2016) Airfoil Data Set. Illinois, USA: University of Illinois at Urbana–Champaign (UIUC), Applied Aerodynamics Group
32. NREL (2012) PreComp: National laboratory of the U.S. Department of Energy, Office of Energy Efficiency and Renewable Energy. Retrieved from https://nwtc.nrel.gov/PreComp
33. Segovia F, Salvador MD, Sahuquillo O, Vicente A (2007) Effects of long-term exposure on e-glass composite material subjected to stress corrosion in a saline medium. J Composite Mater 41(17):2119–2128. doi:10.1177/0021998307074134
34. Bir GS (2006) User's Guide to PreComp (Pre-Processor for Computing Composite Blade Properties). Colorado, USA: National Renewable Energy Laboratory (NREL)
35. Barltrop N, Varyani K, Grant A, Clelland D, Pham X (2006) Wave-current interactions in marine current turbines. Proc Institut Mech Eng, Part M: J Eng Marit Environ 220(4):195–203
36. Gamstedt EK, Sjögren BA (2002) An experimental investigation of the sequence effect in block amplitude loading of cross-ply composite laminates. Int J Fatigue 24(2–4):437–446. doi:10.1016/S0142-1123(01)00099-8
37. Kennedy CR, Brádaigh CMÓ, Leen SB (2013) A multiaxial fatigue damage model for fibre reinforced polymer composites. Composite Struct 106:201–210
38. Purnell P, Cain J, Van Itterbeeck P, Lesko J (2008) Service life modelling of fibre composites: a unified approach. Composites Sci. Technol 68(15):3330–3336
39. Davies P, Mazeas F, Casari P (2001) Seawater aging of glass reinforced composites: shear behaviour and damage modelling. J Compos Mater 35(15):1343–1372
40. Narasimha Murthy HN, Sreejith M, Krishna M, Sharma SC, Sheshadri TS (2010) Seawater durability of epoxy/vinyl ester reinforced with glass/carbon composites. J Reinf Plast Compos 29(10):1491–1499. doi:10.1177/0731684409335451
41. Whitby B, Ugalde-Loo CE (2014) Performance of pitch and stall regulated tidal stream turbines. IEEE Transa Sustain Energy 5(1):64–72. doi:10.1109/TSTE.2013.2272653
42. Carr R (July, 2010) [Tide height data. (personal communication)]
43. SIMULIA (2010) "Abaqus FEA" Dassault Systèmes. from http://www.3ds.com/products-services/simulia/products/
44. Crawford RJ (1998) Plastics engineering: Butterworth-Heinemann
45. Bertagnolio F, Sørensen NN, Johansen J, Fuglsang P (2001) Wind turbine airfoil catalogue. Riso National Laboratory, Roskilde, Denmark

Marine Ageing Behaviour of New Environmentally Friendly Composites

Peter Davies, Pierre-Yves Le Gac, Maelenn Le Gall and Mael Arhant

Abstract In recent years several new materials have been proposed for marine applications. These include liquid infusible acrylics, basalt fibre and plant fibre-reinforced composites, and thermoplastic polyamide composites. In order to assess the long-term durability of such materials accelerated tests are used but the validity of this approach, widely accepted for traditional marine composites, must be checked. This presentation will describe results from ageing tests on these four materials, specifically developed for particular applications: acrylic composites and basalt and flax fibre composites for surface structures, and carbon-reinforced polyamides for deep sea pressure vessels.

Keywords Ageing · Durability · Seawater · Environment

1 Introduction

There is now an extensive database of properties of marine composites and several recent publications provide an overview [1, 2]. Predicting long-term behaviour of complex, anisotropic materials is not simple; while water diffusion can be accelerated by increasing the temperature this can introduce additional degradation mechanisms which may not occur at lower temperatures. In an ideal situation the chemistry of the matrix reactions and the response of the fibre/matrix interface would be fully understood so that the choice of accelerating temperature could be justified. In practice, matrix resins are complex formulations and the large number which are commercially available preclude detailed investigation, so lifetime predictions are often based on experience, a limited number of tests, and safety coefficients.

There has been a growing awareness in the last few years of the need to include environmental impact in the design process. This covers both material selection,

P. Davies (✉) · P.-Y. Le Gac · M. Le Gall · M. Arhant
Marine Structures Laboratory, IFREMER Bretagne Centre, 29280 Plouzané, France
e-mail: peter.davies@ifremer.fr

P. Davies and Y.D.S. Rajapakse (eds.), *Durability of Composites in a Marine Environment 2*, Solid Mechanics and Its Applications 245,
https://doi.org/10.1007/978-3-319-65145-3_12

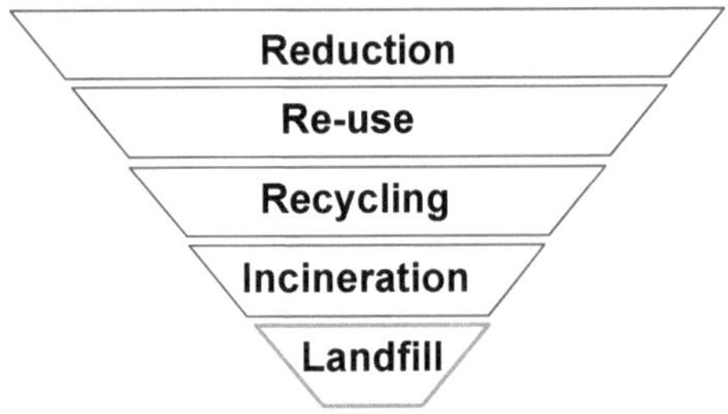

Fig. 1 End-of-life options for marine composites, bottom to top: least to most desirable, based on [4]

in terms of evaluating energy requirements, and also considering end-of-life options. Life cycle analysis provides a standard framework to assess these parameters [3] and has been used in some cases, but input data are not always available. One way to represent the end-of-life options is shown in Fig. 1 [4].

The worst scenario for a structure such as a composite boat is putting components in landfill at the end of their life, as neither materials nor energy is recovered. The materials used today in most marine structures, mainly glass fibre-reinforced polyester, are very difficult to recycle and most will end up at landfill sites. A slightly improved scenario involves incineration, to recover at least some energy, but this requires separation of the many different materials which make up most marine structures. There have been some initiatives to dismantle pleasure boats and in France several sites now exist for this [5]. However, it is clear that we are losing significant amounts of valuable material resources every year, and these resources are limited. Solutions must be found to reduce this waste.

In order to address this, various new materials are being proposed. One option is to develop low-cost thermoplastic matrix composites, as these can, at least in theory, be recycled at the end of their useful life to produce new structures. Another active area concerns new fibre options, and two examples will be presented, basalt and natural fibres from plants. However, in order to transfer all these developments to industrial applications it is essential to evaluate their long-term durability in seawater compared with existing materials; this paper will describe recent work in this area at Ifremer.

2 Materials & Methods

The materials studied here are presented in Table 1, together with material details.

The main equipment used to examine the durability of these materials is a set of natural sea water tanks controlled at temperatures from 4 to 90 °C, with continuous water renewal, Fig. 2. These allow samples to be immersed for long periods, and periodic weighing on a precision balance provides an indication of water diffusion kinetics.

Quasi-static mechanical tests were performed on matrix resins and composites in tension at a crosshead displacement rate of 2 mm/min. An extensometer with a

Table 1 Materials studied

Material	Fibres	Matrix	Fibre Content, V_f	Thickness (mm)	Manufacture
Acrylic	Unreinforced	Elium 150	0	2.7	Cast
Glass/acrylic	E-glass 0/90°	Elium RT300	0.58	2.9	Infusion
Carbon/acrylic	Carbon 0/90, QX	Elium 190	0.52, 0.54	1.7, 1.5	Infusion
Basalt/epoxy	UD	Araldite 1564	0.44	3.5	Infusion
Glass/epoxy	UD	LY/Aradur 3687		3.5	Infusion
Flax/polyester	Biaxial Flax	Bio-sourced polyester	0.3	4	Infusion
Polyamide	Unreinforced	PA6	0	70 µm, 2	Extrusion
Carbon/Polyamide	UD	PA6	0.48	2	Thermo-compression

QX Quadriaxial, 0/45/90°. *UD* Unidirectional

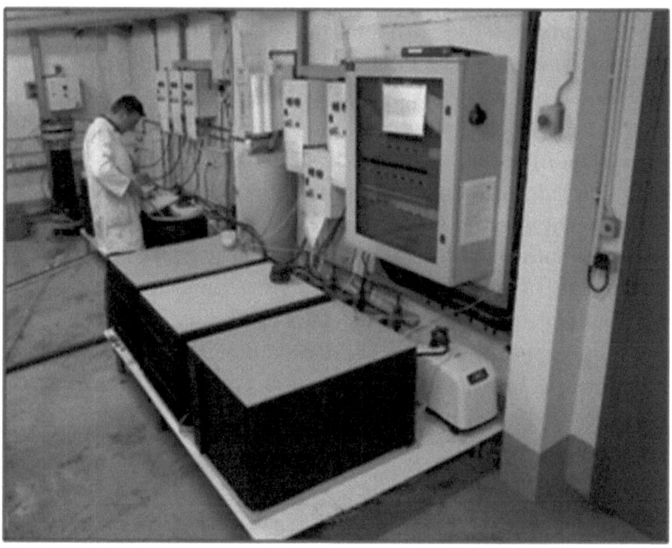

Fig. 2 Natural seawater ageing tanks maintained at different temperatures and continuously renewed

gauge length of 50 mm was used to measure strain. Compression tests were performed using the end pinned buckling test [6] at a crosshead displacement rate of 6 mm/min.

Cyclic tests were performed in four-point flexure, with 100 mm between lower supports, under sinusoidal load control with an R-ratio of 0.1 at 2 Hz.

3 Infusable Acrylic Matrix Composites

Vacuum-assisted infusion is one of the most common manufacturing methods for boat structures and has largely replaced hand lay-up in recent years. This technique is used to produce glass fibre-reinforced polyester, vinyl ester and epoxy structures but these are all thermosetting resins. The recent development of Elium™ acrylic matrix resins offers new options, as these can be manufactured using infusion but polymerize to become thermoplastic, so recycling is possible. However, due to the recent introduction of these materials (2013) very few ageing data are available; a campaign was therefore launched in 2014 to evaluate the sea water ageing resistance of the resin and its carbon and glass fibre-reinforced composites. Coupons and specimens were aged in seawater at 25, 40 and 60 °C for up to 18 months. Figure 3 shows one example of results for mechanical properties of the acrylic matrix resin. Although the initial tensile strength is lower (55 MPa compared to 74 MPa for the epoxy) the strength retention is better than that of this commercial epoxy resin commonly used in marine composites. Final tensile strength after saturation is around 45 MPa, compared to 35 MPa for the epoxy. Modulus values are similar for both before immersion, around 3 GPa, and drop after saturation to 2.5 and 1.9 GPa, respectively, for acrylic and epoxy.

Following these encouraging results, an 18-month ageing study was performed on glass and carbon fibre-reinforced acrylic composites. Figure 4 shows an example of stress–strain plots for carbon/acrylic samples dry, fully saturated after immersion for 12 months in sea water at 60 °C, and dried again after ageing. There is clearly an effect of water but it is at least partly reversible.

These first results suggest that these acrylic materials are quite stable in a marine environment. Further results and more details can be found in [6]. From an ageing point of view this appears to be a good candidate resin for composites used in marines structures; as an example acrylic resin composites were used recently to manufacture a prototype racing yacht [7]. More work is needed now to evaluate its recycling potential.

Fig. 3 Example of acrylic ageing results at 60 °C in sea water, strength loss compared to marine epoxy

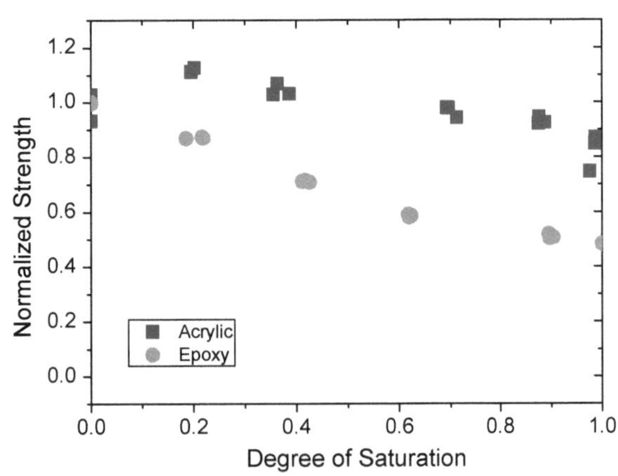

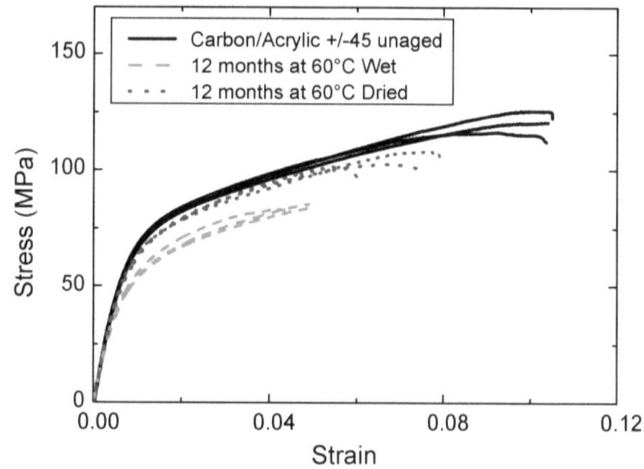

Fig. 4 Influence of immersion on in-plane shear behaviour of ± 45° carbon/acrylic laminates

4 Basalt Fibre Composites

The marine composite industry is dominated by glass fibre-reinforced materials. These have been in use for over 50 years, and provide excellent properties at reasonable cost. For high performance applications, carbon and occasionally aramid fibres are used but there has been no challenge to glass fibres since their introduction. In recent years, however, there has been some interest in basalt fibres for composite structures [8]. Basalt is a natural material found in volcanic rocks; basalt fibres were developed in Russia in the 1950s and are now commercially available. Various reasons have been given for adopting them as a composite reinforcement in place of glass, including lower energy required to manufacture them [9], improved tensile properties, improved corrosion resistance compared to E-glass [10] and improved temperature stability [11]. All these advantages are open to discussion, but if basalt fibre composites are to be used in marine structures it is essential to know how they respond to sea water ageing. Few data are available [12].

A study was therefore performed in which the same epoxy resin was reinforced with either glass or basalt fibres and aged in natural seawater to saturation. Figure 5 shows examples of the diffusion plots. The weight gains and diffusion kinetics are very similar for both materials and governed by the matrix resin.

The initial quasi-static flexural properties of the basalt composites were higher than those of the glass composite, 698 MPa compared to 594 MPa, but in both the dry and saturated states it is interesting to note that the normalised flexural fatigue performance of both materials with respect to quasi-static properties in the same condition (526 and 559 MPa for basalt and glass after ageing) is very similar, Fig. 6.

These results suggest that there is no fundamental difference in the ageing behaviour of basalt fibre composites compared to glass fibre-reinforced materials.

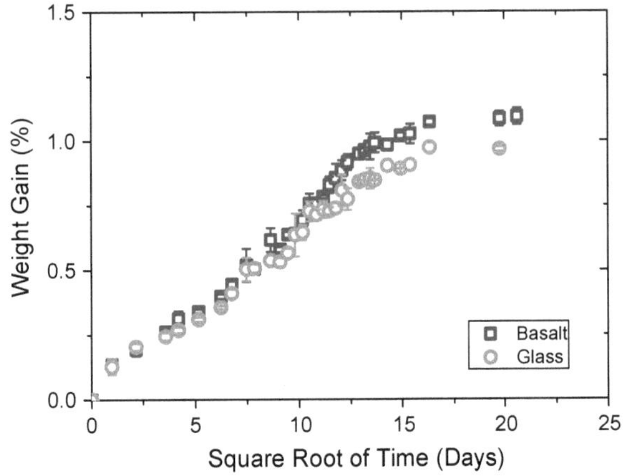

Fig. 5 Weight gains for basalt/epoxy and glass/epoxy at 25 °C

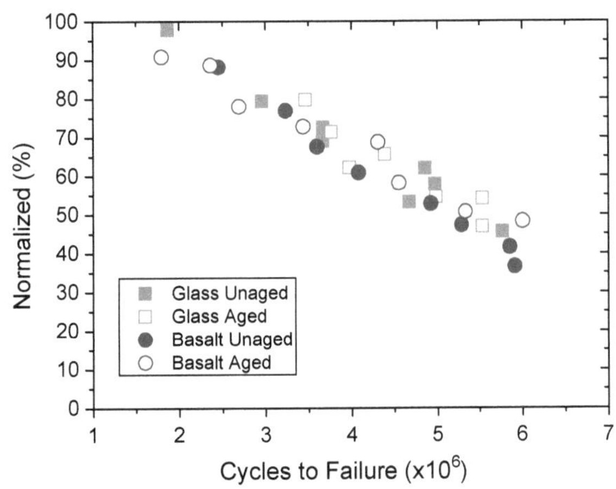

Fig. 6 Normalized flexural fatigue plots, basalt and glass fibre composites, before and after ageing to saturation

5 Natural Fibre Composites

There has been considerable interest in natural fibre composites in recent years, particularly from the automotive industry. These bio-sourced reinforcements require little energy to produce and can be composted at the end of their useful life, providing significantly lower environmental impact than their glass fibre-reinforced equivalent [13]. It is therefore of interest to examine whether they can be used in marine structures. However, the cellulose-based fibres such as flax are very sensitive to water, so the role of the matrix resin is critical in protecting them from water ingress. This is quite different from the case of glass- or carbon-reinforced composites where it is the matrix resin which absorbs water while the fibres absorb very little.

Fig. 7 Ageing setup for one-sided exposure of 70 × 70 mm^2 samples to seawater

Fig. 8 Weight gains for immersion and one-side exposure of flax/polyester, 20 °C

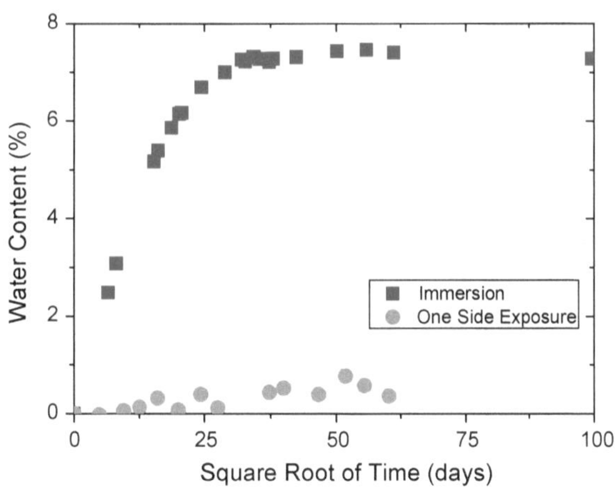

If traditional coupon immersion is used to evaluate the durability of these materials the exposure of fibres at the (unprotected) coupon edges results in very high weight gains and property losses. However, when a more realistic test is performed, in which water only enters through one exposed face, Fig. 7, weight gains are very low, Fig. 8.

These results are quite dependent on the boundary conditions, and in particular the environment in the room where the test cell is placed.

Following an extensive test campaign a flax fibre-reinforced polyester boat was designed and built, the Gwalaz, Fig. 9 [14].

Fig. 9 Gwalaz all-flax reinforcement trimaran

Samples taken from the Gwalaz hull after 18 months' navigation confirmed that very small amounts of water had entered the hull composites, and suggest that ageing using the one-face exposure method is a more realistic test for these materials.

6 Thermoplastic Polyamide Composites

The fourth type of material is being considered for future use in underwater applications, carbon fibre-reinforced thermoplastic. There has been considerable work on high performance thermoplastic matrix polymers such as polyether ether ketone (PEEK), including hydrostatic pressure testing of carbon/PEEK composite cylinders [15, 16]. These materials show excellent properties, but they are very expensive and require high forming temperatures (above 350 °C).

The question addressed here is whether cheaper, lower performance polymers such as polyamides might provide adequate properties for underwater applications. There are many different polyamide grades available (e.g. PA6, PA66, PA11, PA12); one of the cheapest is polyamide 6, but this is known to be very sensitive to moisture. The first step is therefore to study the diffusion kinetics for PA6 in seawater. Figure 10 shows the weight gain of an unreinforced PA6 immersed at 15 °C. This shows that the polymer can pick up 10% by weight of water. This water has a significant effect on properties, but after drying the initial properties are completely recovered [17].

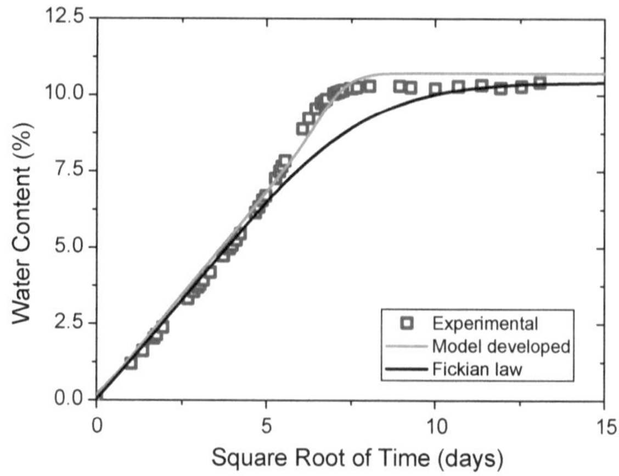

Fig. 10 Weight gain of polyamide 6 immersed in sea water at 15 °C

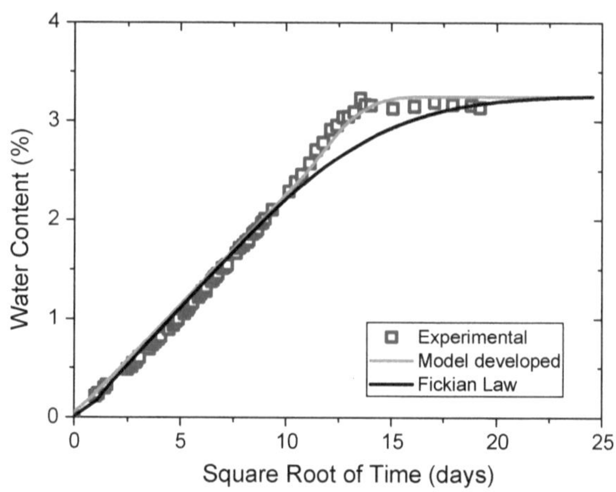

Fig. 11 Composite weight gain, immersion sea water 15 °C

Diffusion studies of this material pose problems, as the dry PA6 glass transition temperature is quite low, around 60 °C. When water enters the material the T_g drops significantly, and during the ageing period the material changes from the glassy to the rubbery state. This results in an acceleration of water ingress as shown in Fig. 10, and the need for a specific diffusion model. One has been developed and more details of this are given in [18].

Underwater pressure vessels are quite thick, in order to resist hydrostatic pressure, so it is also important to know the composite diffusion kinetics, as in service water ingress may be confined to a small external region of the wall thickness. Immersion tests have therefore been performed on composites, Fig. 11. Here saturation occurs at around 3% weight gain due to the presence of fibres, and again the non-Fickian behaviour can be observed.

Hydrostatic pressure may also affect water ingress [19], so a series of tests was performed under pressure, Fig. 12. Samples were placed in water in a pressure

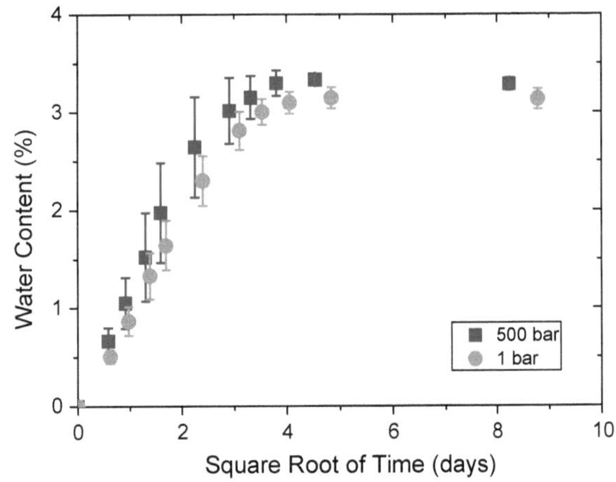

Fig. 12 Influence of pressure on weight gain, carbon/PA6, 40 °C

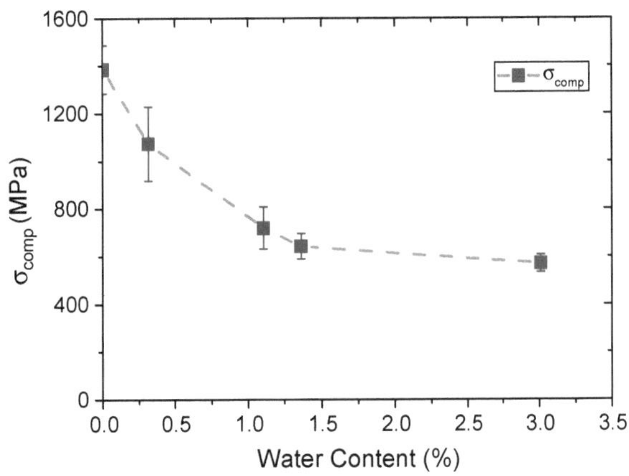

Fig. 13 Influence of water content on compression strength, saturated carbon/PA6 specimens

vessel inside an oven at 40 °C. Periodic weight measurements allow plots of water content versus square root of time to be constructed. Results for samples subjected to pressure are compared to those for similar specimens in the same oven without pressure. These show that there is not a significant influence of pressure on diffusion in this material.

The influence of moisture content on composite properties was then examined. Figure 13 shows one example, the influence of water on axial compression strength measured on unidirectional specimens using the end hinged buckling test. Valid compression strength values are hard to obtain, and many test variants exist. This test involves flexural loading of a long specimen and allows compression properties to be obtained by strain gauging specimens subjected to axial buckling. More details can be found in [20]. Higher values are obtained than for pure compression due to the presence of a strain gradient across the specimen thickness [21]. Tests were performed on specimens which had been saturated in water at different

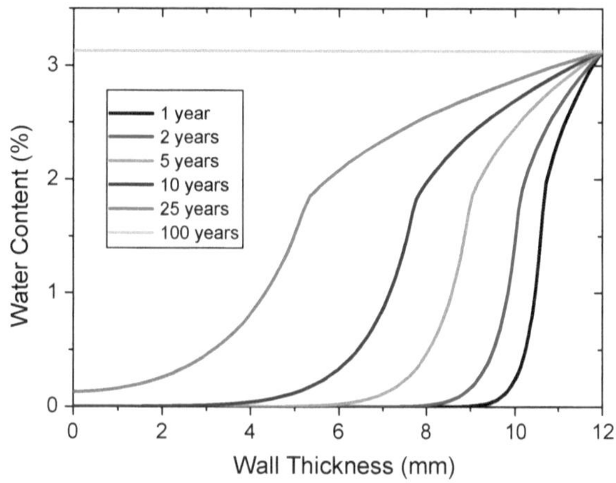

Fig. 14 Predicted water content profile in 12-mm-thick tube

saturation levels (by exposing specimens to different humidity conditions using salt solutions), in order to avoid the presence of a water gradient. These data are very useful as they can be introduced into an ageing model for any geometry once the local water profile has been established.

There is indeed a drop in compressive strength with water ingress, and all the matrix-dependent properties decrease, but the changes are predictable and residual strength values around 600 MPa are measured at complete saturation. The properties are also recoverable after drying, indicating that matrix plasticization is the main physical mechanism involved.

Based on the water diffusion kinetics and the loss in compression strength, it is possible to estimate the response of a composite tube in deep sea. For example, for a 12-mm-thick carbon/polyamide 6 pressure vessel of 120 mm inner diameter at 15 °C it would take around 75 years to saturate the wall thickness, Fig. 14, so the use of these materials for deep sea applications such as drifting profilers (2 years' immersion at 2500 m depth) appears quite feasible. More details are available in [22, 23].

7 Conclusions

This paper describes the tests used to evaluate the durability of composite materials developed to reduce environmental impact of marine structures. For the first, acrylic matrix composites, the traditional approach with ageing accelerated by raising water temperature, are satisfactory. Results show that these materials are quite stable in seawater.

For the second, basalt fibre composites, a classical study comparing weight gain and property changes for basalt fibre with those for glass fibres in the same epoxy matrix, showed property retention at least as good as for glass. This suggests that basalt fibres may be used to reinforce marine structures if their environmental advantages are confirmed.

For the third material type, natural fibre-reinforced biocomposites, the usual immersion of coupons leads to very high weight gains and property losses due to the hydrophilic nature of cellulose-based fibres and diffusion dominated by simple edge effects. An alternative approach for these materials has been proposed, using exposure to only one surface.

And for the final case, thick carbon/polyamide 6, the low Tg of PA6 causes a transition from the glassy to rubbery state during immersion. This requires a specific analysis, but the diffusion kinetics are such that despite the high affinity of PA6 for water, the time for saturation in the configuration considered is so long that degradation is quite limited and predictable.

These results suggest that there is considerable potential for reducing environmental impact in many marine applications.

References

1. Weitsman YJ (2012) Fluid effects in polymers and polymeric composites. Springer
2. Davies P, Rajapakse Y, Editors (2013) Durability of composites in a marine environment. Springer
3. ISO 14040 (2006) Environmental management–Life cycle assessment—Principles and framework
4. Azapagic A, Emsley A, Hamerton I (2003) Polymers, the Environment and Sustainable Development. Wiley
5. APER (Association pour la Plaisance Eco-responsable) website. http://www.aper.asso.fr/
6. Davies P, Le Gac PY, Le Gall M (2016) Influence of sea water ageing on the mechanical behaviour of acrylic matrix composites, Appl Composite Mater. on-line, doi:10.1007/s10443-016-9516-1
7. Arkema website. http://www.arkema.com/en/innovation/arkema-sailing/arkema-boats/mini-6-50-prototype/
8. Fiore V, Scalici T, Di Bella G, Valenza A (2015) A review on basalt fibre and its Composites. Compos B 74:74–94
9. Sim J, Park C, Moon DY (2005) Characteristics of basalt fibre as a strengthening material for concrete structures. Compos Part B 36:504–512
10. Nasir V, Karimipour H, Taheri-Behrooz F, Shokrieh MM (2012) Corrosion behavior and crack formation mechanism of basalt fibre in sulphuric acid. Corros Sci 64:1–7
11. Hao LC, Yu WD (2010) Evaluation of thermal protective performance of basalt fibre non-woven fabrics. J Therm Anal Calorim 100:551–555
12. Liu Q, Shaw MT, Parnas RS, McDonnell M (2006) Investigation of basalt fiber composite ageing behavior for applications in transportation. Polymer Composites, 75–483
13. Le Duigou A, Davies P, Baley C (2011) Environmental impact analysis of the production of flax fibres to be used as composite material reinforcement. J Biobased Mater Bioenerg 5(1):1–13
14. Bosser L, Poisson E, Grossmann E, Davies P (2013) The first all-flax composite trimaran, JEC magazine, No.84 October, 45–47
15. Gruber MB, Lamontia MA, Smoot MA, Peros V (1995) Buckling performance of hydrostatic compression-loaded 7-inch diameter thermoplastic composite monocoque cylinders. J Thermoplast Compos Mater 8(1):94–108

16. Davies P, Riou L, Mazeas F, Warnier P (2005) Thermoplastic composite cylinders for underwater applications. J Thermoplast Composite Mater 18(5):417–431
17. Le Gac PY, Arhant M, Le Gall M, Davies P (2017) Yield stress changes induced by water in polyamide 6: characterization and modeling, Polymer Degradation and Stability, In Press
18. Arhant M, Le Gac PY, Le all M, Burtin C, Davies P (2016) Modelling the non Fickian water absorption in polyamide 6. Polym Degrad Stab 133:404–412
19. Humeau C, Davies P, Jacquemin F (2016) Moisture diffusion under hydrostatic pressure in composites. Mater Des 96:90–98
20. Fukuda H (1989) A new bending test method of advanced composites. Exp Mech 29(3): 330–335
21. Wisnom MR, Atkinson JW (1997) Constrained buckling tests show increasing compressive strain to failure with increasing strain gradient. Compos A 28(11):959–964
22. Arhant M, Davies P, Burtin C, Briançon C (2015) Thermoplastic matrix composites for underwater applications. In: Proceedings 20th international conference on composite materials, (ICCM20), Copenhagen
23. Arhant M, Le Gac PY, LeGall M, Burtin C, Briancon C, Davies P (2016) Effect of seawater and humidity on the tensile and compressive properties of carbon-polyamide 6 laminates. Composites Part A: Appl Sci Manuf 91(1):250–261

Erratum to: Durability of Composites in a Marine Environment 2

Peter Davies and Yapa D.S. Rajapakse

Erratum to: P. Davies and Y.D.S. Rajapakse (eds.), *Durability of Composites in a Marine Environment 2*, Solid Mechanics and Its Applications 245, https://doi.org/10.1007/978-3-319-65145-3

In the original version of the book, series volume number has been changed from 244 to 245. The erratum book has been updated with the change.

The updated online version of this book can be found at
https://doi.org/10.1007/978-3-319-65145-3

P. Davies and Y.D.S. Rajapakse (eds.), *Durability of Composites in a Marine Environment 2*, Solid Mechanics and Its Applications 245,
https://doi.org/10.1007/978-3-319-65145-3_13

MIX
Papier aus verantwortungsvollen Quellen
Paper from responsible sources
FSC® C105338

If you have any concerns about our products,
you can contact us on
ProductSafety@springernature.com

In case Publisher is established outside the EU,
the EU authorized representative is:
Springer Nature Customer Service Center GmbH
Europaplatz 3, 69115 Heidelberg, Germany

Printed by Libri Plureos GmbH
in Hamburg, Germany